A Human History
of Emotion

A Human History
of Emotion

How the Way We Feel
Built the World We Know

RICHARD FIRTH-GODBEHERE

4th ESTATE • *London*

4th Estate
An imprint of HarperCollins*Publishers*
1 London Bridge Street
London SE1 9GF

www.4thestate.co.uk

HarperCollins*Publishers*
1st Floor, Watermarque Building, Ringsend Road
Dublin 4, Ireland

First published in Great Britain in 2021 by 4th Estate
First published in the United States by Little, Brown Spark in 2021

1

Set in Janson Text
Printed and bound in the UK using 100%
renewable electricity at CPI Group (UK) Ltd

MIX
Paper from
responsible sources
FSC
www.fsc.org
FSC™ C007454

This book is produced from independently certified FSC™ paper
to ensure responsible forest management.

For more information visit: www.harpercollins.co.uk/green

This book is dedicated to my two late dads,
Raymond Godbehere and Roger Hart.
I think you'd both have gotten a kick out of this.

Contents

A Human History
of Emotion

Introduction

How Do You Feel?

My cat spends a lot of her time being angry. Her usual way of displaying this rage is by tapping and chasing her tail while shrieking, snarling, and hissing. An outside observer might think she simply hates her tail, but I assure you it's a display of bad temper, and it's aimed at me. She does it when I feed her half an hour too late, or sit in her spot on the sofa, or commit the heinous crime of letting it rain. Of course, Zazzy is far from the only pet to express fury at the disobedience of her owner. Anyone with a cat, dog, rabbit, snake, or whatever knows that pets feel emotions and express them every chance they get. They can be angry, demanding, and loving — often at the same time. Emotions seem to flow through our animal companions just as freely as they do through us.

But here's the twist: pets don't feel emotions. And before you find yourselves defending a hill marked MY CAT LOVES ME, it's not just pets. Humans don't feel emotions, either. Emotions are just a bunch of feelings that English-speaking Westerners put in a box around two hundred years ago. Emotions are a modern idea — a cultural construction. The notion that feelings are something that happens in the brain was invented in the early nineteenth century.[1]

According to linguist Anna Wierzbicka, there's only one word about feelings that's directly translatable from one language to another: *feel*.[2] But what you can *feel* goes far beyond what's usually thought of as emotion: there's physical pain, hunger, warmth or cold, and the

3

sensation of touching something. In the English language alone, various terms have been used at various points in history to describe certain types of feelings. We've had *temperaments* (the way people's feelings make them behave), *passions* (feelings felt first in the body that affect the soul), and *sentiments* (the feelings you get when you see something beautiful or someone acting immorally). We've left most of these historical ideas behind, replacing them with a single catchall term that describes a certain type of feeling processed in the brain: *emotion*. The problem is that it's difficult to pin down the types of feelings that do and don't constitute emotions. There are almost as many definitions of emotions as there are people studying them. Some include hunger and physical pain; others don't. The concept of "emotions" isn't true, and the notion of "passions" isn't false. "Emotion" is just a newer box. A box with poorly defined edges, I might add. The question, then, is if emotion is really just some vague modern construct, where do you begin when writing a book about it?

What Is Emotion?

The biggest problem with trying to answer the question "What is emotion?" is that it's a bit like trying to answer the question "What is blue?" You might be able to point to some scientific data about light refraction and wavelengths, but the fact is that blue means many things to many people. Some cultures, such as the Himba tribe of Namibia, don't recognize blue as a color at all. They think of it as a type of green, one of many greens that allow them to differentiate among the subtle hues of the leaves in the jungles and grasslands they live in. Knowing a safe blue-green leaf from a poisonous yellow-green leaf could mean the difference between life and death.[3]

If we devised a test for colors and asked members of the Himba tribe to sort objects that look like the color of grass into one pile and objects that look like the color of the sky into another, you'd get one pile with lots of greens and another with lots of blues. That might, understandably, make you think that the concepts of green and blue are universal. But if, instead, you asked them to sort the objects into a blue pile and a green pile, you might see a whole heap of blue things in what a Westerner would likely call the green pile. Then, just as understandably, you'd think that the perception of color is culturally constructed.[4]

Similarly, you might take pictures of people making faces based on emotions as you understand them, then ask the equivalent of "What color is the sky?" For example, you might ask, "What face would you make when you eat something rotten?" Then when the tribe pointed to a picture of a "gape face" (the downturned, slightly open-mouthed, wrinkled-nose-and-narrow-eyed expression many in the West associate with disgust), you'd be justified in claiming that disgust is universal. Alternatively, you could take pictures of a range of facial expressions and ask a group of people to sort them into a "disgust" pile and an "anger" pile. You might be surprised to find the gape face in the anger pile along with expressions of surprise, anger, fear, and confusion. If that happened, you might become convinced that emotions are culturally constructed. The question is, Which of these methods is right? Is it nurture or nature? Well, as so often is the case when it comes to similar one-way-or-the-other questions, the answer is probably yes.

I'll get into this in much more detail later in the book, but for now, suffice it to say that both culture and biology matter. We are taught how we are supposed to behave when we feel something by our upbringing and our culture. But our feelings themselves may share an evolutionary origin. In the same way that the Himbas' understanding of the color green differs from mine, context, language, and other cultural factors play a role in the way a human understands emotions. We might all feel similar things, but the way we understand and express those feelings changes from time to time and from culture to culture.

Those important differences are where the history of emotion, and this book, live.

What Is the History of Emotion?

I am planting my flag firmly in a growing discipline called the history of emotion. It's a field that tries to understand how people understood their feelings in the past. Some studies sweep across huge expanses of time, examining the long history of human fear.[5] Others are quite specific, exploring the ways emotions were understood in small geographical areas during particular periods—for example, a study of the emotional regimes at play in the French Revolution.[6] (I'll discuss emotional regimes in a bit.)

The history of emotion is a discipline that has thrown up hundreds of theories and ideas, and it's having an increasing impact on the way we understand the past. But most of the work in this field has been niche and academic—not the sort of thing you might want to read while relaxing at the seaside. I've written this book because I'm on a bit of a mission to share the wonderful world of the history of emotion with as many people as I can—to let as many people as possible share the excitement and perspective offered by this new way of understanding times gone by and to provide a new way for people to see the world, particularly its past.

There are hundreds of ways that you can study emotions in history. You can write material histories of objects that tell emotional stories, such as scented letters, religious artifacts, and children's toys.[7] You can examine how the names of emotions have changed over time and how words for describing emotions have shifted in meaning. For example, the English word *disgust* once just meant "things that taste bad." Now it refers to an aversion to anything repulsive—from moldy fruit to rotten behavior.[8] Sometimes, the history of emotion is a bit like the field of intellectual history or the history of ideas and science in that it strives to discover what people once thought feelings were and how they understood emotions within the context of their times and cultures. There are many ways in, but there are a few frameworks that we historians of emotion keep coming back to, regardless of subdiscipline.

The first of these is the one I mentioned above: *emotional regimes.* This term, coined by historian William Reddy, denotes the expected emotional behaviors imposed on us by the society we live in. These regimes attempt to explain the ways emotions are expressed in any given set of circumstances.[9] For example, an airline steward in the first-class compartment is generally expected to be polite and courteous to passengers, no matter how rude they are. His job imposes an emotional regime on him that soon becomes second nature: polite calmness and unending patience.

Linked closely to emotional regimes is something called *emotional labor.* This term has been expanded to mean almost anything, from merely being polite to being the person in the household (usually a woman) who performs emotion-related tasks, such as sending birthday cards and keeping the house clean to impress the visiting neighbors. But it initially had its roots in Marxist thought. The term was coined by sociologist Arlic Hochschild. She described emotional labor as the need to "induce or suppress feelings to sustain the outward countenance that produces the proper state of mind in others."[10] This may sound a bit like an emotional regime. The difference, as another sociologist, Dmitri Shalin, put it, is that emotional labor is "the emotional surplus meaning systematically extracted by the state [or its emotional regime] from its members." To go back to our airline steward, the emotional regime is what keeps him smiling even when the customer is rude. The emotional labor is the effort it takes to keep smiling, even though deep down inside he wants to scream at the customer. In other words, emotional labor is the effort required to stay within an emotional regime. Emotional labor exists because emotional regimes are top-down, imposed from some sort of higher authority, often the state but sometimes religions, philosophical beliefs, or the moral codes we are raised to adhere to.

Because emotional labor can be physically and mentally tiring, staying true to an emotional regime is hard. People need access to venues where they can let their emotions out. William Reddy coined the term *emotional refuges* to describe them. The hotel bar that the airline steward frequents to vent to his colleagues about the rude man in first class might be one such refuge. These refuges can be engines of revolution, especially when the suppressed feelings become the fuel for a change in emotional regime.

But the way we express our emotions isn't always imposed on us from above. Sometimes, it rises up from people and culture. These bottom-up emotional rules are known to historians of emotion as *emotional communities*, an idea first suggested by historian Barbara Rosenwein.[11] It refers to the currents of shared feeling that bind a community together. If you have ever visited your in-laws for a seemingly endless hour or so, you'll know what I mean. The ways they express themselves can be quite different from the ones to which you've become accustomed. For example, my family is quite boisterous. We—and this includes my mom—like rude jokes, silly stories, gently teasing each other, and, because we are a mostly academic family, delivering highbrow conversation in as lowbrow a way as possible. I wouldn't dream of inflicting that sort of behavior on my wife's family. That's because each family has formed its own emotional community, its rules for behavior and expression.

You get the same feeling when traveling to other countries. In fact, you don't have to travel very far. I've been to concerts in Barnsley, England, where the audience remained stony-faced and motionless during the performance. But after the music stopped, there was a line of people ready to go over to the band, buy them beers, and tell them how great they thought they were. That town's specific emotional community is one where stoic manliness—regardless of gender—

denies the sort of passionate expression you see at performances elsewhere, even in towns a few short miles away.

People can be part of more than one emotional community or regime. For example, our airline steward's tolerance for his job's emotional regime doesn't extend to his soccer fandom. While standing on the terraces among his Manchester United kinfolk, the man who displays seemingly endless patience at work can be savage and rude to a fan of an opposing team. While he's at the game, he's living in an emotional community and free of the emotional regime that governs his conduct at work. He is free to express his emotions as that community sees fit.

This brings me to another central aspect of this book. Throughout history, certain powerful emotions have acted as a driving force for change. On many occasions, desire, disgust, love, fear, and sometimes anger seem to take over cultures, making people do things that can change everything. I'll explore how these emotions—and people's evolving conceptions of them—played a role in shaping the world. In the process, we'll see how people's experience of desire, disgust, love, fear, and anger in the past was different from the way we experience these emotions today.

What follows is a sweeping tour through the various ways people have understood their feelings through the ages, one that will illustrate how feelings have changed the world in ways that still reverberate today. We'll cover everything from the ancient Greeks to artificial intelligence, traveling from the shore of Gambia to the islands of Japan, to the might of the Ottoman Empire, to the rise of the United States. We'll even take a peek into the future.

History shows us that emotions are powerful—that they, as much as any technology, political movement, or thinker, built the world. They laid the foundations of religions, philosophical inquiries, and the pursuit of knowledge and wealth. But they can also be a dark force, capable of tearing worlds down through war, greed, and mistrust.

Each of the following chapters focuses on a specific time and place, but taken together they provide a narrative of how emotions shaped the world we live in today in all its complexity, wonder, and diversity. I hope that by the end you'll agree, and you'll never think about emotions in the same way again.

One

Classical Virtue Signaling

Let's begin with some big ideas. History is full of ideas about emotions—what they are, where they came from, how they ought to be expressed and controlled. These ideas helped form the religions and philosophies that are still with us today. In many cases, ideas about feelings had an impact that shaped history. But before we get to the chapters about ancient India, the New Testament era, and the ideas of saints and prophets, I'm going to start at the beginning—or at least the beginning we know about, which gave rise to some of the first ideas about emotions on record. It means that, as often is the case, we need to travel back to ancient Greece.

Plato and Socrates

Roughly 399 years before Jesus was born, a twentysomething man lay ill in bed.[1] His stocky physique was well known in Athens; it had helped him become a wrestler of quite some fame. He may even have competed at an Olympics. Most of us know him by his nickname, Broad—or, to use the ancient Greek version, Plato.[2]

Plato was not just physically daunting; he was also an intellectual giant. Later in life, he founded a school so important that its name—the Academy—is still used to this day to describe seats of learning. In his Academy, Plato wrote works of philosophy. But he didn't write long prose. He wrote a series of debates that came to be known as dialogues. In all but one of these, the main speaker was his old tutor, Socrates, whom he loved dearly.

It's hard to overestimate how important these dialogues were. More than two millennia later, the philosopher and mathematician Alfred North Whitehead described all the philosophy that came after them as "a series of footnotes to Plato."[3] But without the events of that deeply emotional day when Plato lay sick in bed in 399 BCE, and those that led up to them, he might have just been another of the hundreds of great thinkers who have been lost to time. Because on the same day that Plato was nursing his illness, Plato's teacher, Socrates, was being executed. Plato's feelings about that were, well, complicated.

Feeling Platonic

The Greeks called emotions *pathē*, meaning "experience" or "suffering." Whether it was one or the other depended on which *pathē* you were experiencing (or suffering). Plato believed that *pathē* were disturbances in our souls, ripples caused by external events or sensations that knock you off balance and disturb your calm. But to Plato, souls were more than just the bit of us that isn't flesh.

Souls were important to Plato because they were the human part of an idea that was central to his philosophy. He didn't think that the world we see around us is all there is. He thought that everything, from humans to trees to chairs, was just an imperfect version of what he called the wise cosmos (*kósmos noetós*), better known as forms. He believed we are all born with an inherent knowledge of these perfect forms. That's why we can recognize that two different objects—say, a tavern stool and a throne—are both fundamentally chairs. Both map onto a remembered form of a perfect chair. Plato likened our experience of reality to the experience of people who live in a cave seeing shadows cast on a wall by events outside. What we think is real is just a shadow. In Plato's view, our souls are the reality—our perfect form dancing in the sunlight at the cave entrance. Our bodies are just the shadows they cast. When we feel *pathē*, it's the result of something perturbing our souls, causing sensations in our bodies and making the shadows twist unexpectedly. What confused Plato was how on earth people could feel two different emotions at once. How could someone feel both terrified and brave at the same time, desiring to fight while wanting to flee, for example, like soldiers in a battle? The answer he arrived at was that we have more than one part to our souls.

He reasoned that because animals have souls but can't think in complex ways, there must be one type of soul for animals and another type for humans and gods. The god soul was pure reason and couldn't be perturbed by *pathē* directly. He called this soul the *lógos*.[4]

Lógos is a hard word to translate. It means "thought" or "word," or perhaps the ability to form words into thoughts. Most important, it has a divine element. A useful illustration of this concept appears in John 1:1, in the New Testament. Originally written in Greek, it says (in the King James Version), "In the beginning was the Word [*lógos*], and the Word [*lógos*] was with God, and the Word [*lógos*] was God." If you've ever wondered how God could be described as a word, you may (understandably) be taking this passage a bit too literally. God, here, is really being described as a thought, a soul of pure reason, an ability to know things. That was Plato's *lógos*—a type of soul that can reason, know, understand.

Plato called the soul that animals have the *epithumêtikon*, a word that means "desiring" or "appetitive."[5] When this soul is perturbed by *pathē*, it creates the basic drives that get you through day-to-day life—pleasure, pain, the desire for food and sex, the wish to avoid harmful things, and so on. Because humans are part animal, but obviously capable of more complex reasoning, knowledge, and understanding,

Plato thought we must have both the rational *lógos* and the irrational *epithumêtikon*.

But there must be another part of our souls, too, thought Plato. Humans can feel what is good and bad and act accordingly without having to think about it. Pure logic doesn't do that, nor do our animal appetites, so there must be a third part of the soul. He called this third part of the soul *thumoeides*, or *thymos*—the "spirited soul."[6] *Thymos* translates as "anger," and it's in this part of the soul where you find the feelings that get stuff done. Like the *epithumêtikon*, the *thymos* can be disturbed directly by *pathē*. When the *thymos* is perturbed, it creates anger, obviously. But that sort of perturbation can also cause the *pathē* of "hope," which gets you to do things because you think they might be possible, even if they're difficult. It can create the suffering of "fear," which helps you escape from dangerous situations you were unable to avoid. Or it can induce the experience or suffering of "courage," which gets you to do things even when you're frightened. But—and Plato thought this was very important—the goals the spirited soul aims toward are not necessarily for the greater good. These *pathē*, like the animal soul, make you want to automatically seek pleasure or avoid pain without any thought. This reason-free drive toward pleasure is called *boulesis*. *Boulesis* is not virtuous, because sometimes doing the right thing is painful and doing evil things can give you pleasure.

To be truly virtuous, you need to strive for a type of good that comes from the *lógos*—*eros*. *Eros* isn't about personal pleasure but the greater good. To act virtuously, you can't just let your *pathē* guide you. You have to learn to think about what's really best—to evaluate, to judge. You have to stop and think, "Is this really the right thing to do?" You can't just do it because it gives you nice feelings. The right thing to do might even make you feel bad, taking you away from *boulesis*. But it's still the right thing to do. That is *eros*. The distinction between *boulesis* and *eros* is a vital component of the emotional regime Plato constructed for his readers and followers. It even applied when someone they loved was about to be executed. Plato used the story of Socrates's death as an example of the power of *eros* in the face of *boulesis*. But to get to that story we need to understand why Socrates was put to death in the first place.

The Trial of Socrates

Socrates was convicted of impiety and corruption of the young, and although that's not really why many Athenians wanted him dead, it's hard to argue that he wasn't guilty. He was certainly guilty of corrupting the young. Socrates's tactic, which has come to be known as the Socratic method, involved asking young men questions about their beliefs. Sometimes his questioning challenged the authorities, widely held notions of justice, and even the gods themselves. As Socrates's interlocutors answered, he would ask them more questions, encouraging them to further challenge themselves and refine their ideas. Eventually, the Socratic method would often result in these men convincing themselves that Socrates was right about everything, including his impious ideas.

At the time, Athens had just begun to recover from a century of war and oppression. After a long war with the Persians followed by a bitter civil war with Sparta—during which Socrates became a respected and decorated soldier—the Spartans suspended Athens's famous democracy and installed the Thirty Tyrants in its place. But the Athenians, frustrated by their newly imposed government, soon rebelled. It took them less than a year to boot the Thirty Tyrants out and arrest the people suspected of helping them.

Socrates was one of those arrested. His biggest offense wasn't the impiety or the corruption of the young: it was the matter of whom, exactly, he had been corrupting—many of them were powerful, influential, and deeply hated people. They included Alcibiades, a prominent military general who continually flip-flopped between the Athenian and Spartan armies, depending on which best advantaged him. Socrates's audience also included members of the Thirty Tyrants and the families who supported them. One such person was Critias, one of the most powerful of the Thirty.[7] Another was the son of Critias's niece Perictione: a young wrestler called Plato.

That Socrates's arrest was politically motivated there is no doubt, but he was also guilty of the charges brought against him. Upon being found guilty, Socrates asked that instead of a death sentence the authorities provide him with free meals for the rest of his life in return for his services to the city. That went down about as well as you might imagine, and he was sentenced to death by poison.

The Death of Socrates

The death sentence was carried out when Socrates voluntarily drank a vial of hemlock. According to Plato's account—which he claims to have gotten from another of Socrates's students, Phaedo, who was actually there—when the people who were with Socrates saw him drink the poison, they started crying. Socrates got annoyed, asking, "What is this…you strange fellows. It is mainly for this reason that I sent the women away, to avoid such unseemliness, for I am told one should die in good omened silence. So keep quiet and control yourselves."[8] Their sorrow was born of grief and a need to find a way to change a painful situation. But Plato believed that, as men—and it was exclusively men—they should control themselves. He thought it was fine for women to weep, beat their chests, and tear their tunics. But not men. Their crying was selfish. It was about their selfish aversion to emotional pain and what they *wished* was good, not what *was* good.

After this scolding, the men in the room immediately stopped crying. To resist tears at the death of their friend must have meant investing vast amounts of emotional labor. Still, they felt ashamed of their behavior and realized that they cried not for Socrates—Socrates, it seemed, was content—but rather for their "misfortune in being deprived of such a comrade."[9] In other words, their crying wasn't *virtuous*. It was selfish and therefore ran against the emotional regime that Socrates and Plato prescribed.

There's another part of Plato's account of Socrates's death that perfectly demonstrates his belief about keeping *pathē* in check for the greater good.[10] According to Plato, Socrates was offered the chance to escape.[11] Running away would have felt like the right thing to do. His spirited soul would have been all for it—not dying is undoubtedly good on a personal level. However, he had been tried and found guilty, and that was that. Cheating the law would be wrong, unvirtuous. Plato's Socrates believed that to give in to his feelings would be to turn away from justice, an act that would take him away from *eros* and toward *boulesis*. That would not do in Plato's emotional regime.

According to Plato, Socrates's final words were, "Crito, we owe a cock to Asclepius; make this offering to him and do not forget."[12] There's been a great deal of debate over what this means. Asclepius was the god of healing; surely Socrates didn't think he would be healed of a dose of fatal poison. Some think Socrates was babbling incoherently as the poison took hold.[13] German philosopher Friedrich Nietzsche thought he was saying that "life is an illness," and he was glad to be cured of it.[14] Some think that Socrates was thinking of his young friend Plato, who, let's remember, was supposedly laid up sick in bed.[15] We will probably never know for sure. But I think that perhaps Socrates was thanking Asclepius for healing the city he loved so much. Maybe he knew his execution would act as an emotional release, a catharsis, and that it ultimately served the greater good of Athens. It was the most virtuous act possible. The highest possible example of *eros*. For Plato, feelings had to be controlled for a greater good, so this explanation fits. Plato was using his friend's death to tell us all how the greatest man he ever knew could control his desires and focus on *eros*, even as he was being executed.

The only other account we have of Socrates's death comes from the soldier and follower of Socrates known as Xenophon. He wrote that Socrates was happy to die because, while he was still intellectually sharp at seventy years old, he was worried that he soon wouldn't be.[16] Xenophon's Socrates was a much more practically minded man than Plato's, spending as much time offering advice as arguing. Xenophon's account may be nearer to the truth. But Plato's writings aren't about the truth; he isn't presenting us with facts. He is teaching us a lesson about how the supremely virtuous can control their emotions for the greater good and how we ought to emulate that. He is setting forth an emotional regime—a set of rules for feeling and expressing feelings that he believes we should all abide by.

So what was Plato's emotional regime? To put it as briefly as I can, it was the belief that the greater good is not about giving in to *pathē*— those perturbations of the soul that drive appetites and anger. Nor should you do what you think is right just because it feels right. You must use your *lógos* to find the greater good, the *eros*, in all things, and focus your actions on attaining it. Even if doing so causes your death.

Quite a few people stuck closely to Plato's regime. Some people morphed it into something else—Stoicism—a regime of its own that we'll come to later in the book. Others, like a young follower of Plato who went by the name of Aristotle, rejected Plato's emotional regime almost entirely after coming to conclusions of his own.

Great Expectations

In roughly 334 BCE, sixty-five or so years after Socrates's execution, a young man sat in a tent reading an important letter. According to the Greek biographer Plutarch, who, it must be noted, based his description on statues, this lad was small but muscular and hard. His light pinkish-red face was cleanly shaved, which was unusual for the time, and sat on a bent neck. His head leaned to one side most of the time, making his mismatched blue and brown eyes seem as though they were always looking up at something. It's this physical imperfection that gives some credence to Plutarch's description.[17] The best modern guess as to why his neck tilted in that way is that he had some sort of physical ailment such as congenital muscular torticollis or ocular torticollis.[18] But neither his height, his youth, his shaved chin, nor his neck problems stopped Alexander from becoming "great." By 334 BCE, when Alexander was only twenty-two, he had liberated the Greeks, who were still under the yoke of the Persian forces some seventy years after the end of the Peloponnesian War. But Alexander wanted to go further. He wanted to invade Persia itself. He stood on the border between his kingdom and Asia and threw a spear. If it landed in Persia, he would conquer it. If it didn't, he wouldn't. After seeing the tip pierce Persian soil, he declared that the gods were offering Persia to him as a gift, and he would take it.

According to a story written by a man whom we historians call Pseudo-Callisthenes, the letter Alexander was reading came from a furious King Darius III of Persia. It was a letter filled with boasts and threats.* Darius claimed he was not just a king but also a god, a very rich god, who regarded Alexander as a "servant" who ought to go home and "remain on the lap of [his] mother...for at [his] age, one still requires training and nursing and lap-feeding."[19] This was not a subtle rebuke. A lesser leader might have turned back, worried that his men might find out about the power of the man who would meet them should they march on. But Alexander was smart. He was once tutored by a man whose work is so crucial to the study of philosophy that it's still analyzed and debated today, despite the fact that his books have been lost to history, leaving only his lecture notes. That tutor and erstwhile pupil of Plato: Aristotle.

---

* Bear in mind that there's a reason we call the writer *Pseudo*-Callisthenes. But even though the story may well be made up, it still goes a long way toward explaining Aristotle's emotions.

The Affections of Aristotle

If all Western philosophy is just a series of footnotes to Plato, the notes written by Alexander's old teacher, Aristotle, were more thorough than most. Aristotle was a lot more practical than Plato. Whereas Plato—who often wanted to just sit and have a good think—liked to talk with his friends, Aristotle preferred measuring and looking at the world. This difference of opinion put Plato and Aristotle somewhat at odds. Eventually, in 348 BCE, Aristotle left Plato's Academy. His reasons for doing so aren't clear. Perhaps he had learned all he could from the school. Maybe it was because he didn't get along with Plato's successor, Speusippus. It could have been because Athenians were a little bit xenophobic about Macedonians such as Aristotle. In any case, after spending some time traveling, he returned home and took a job as head of the school at the temple of the nymphs in Mieza, Macedonia. It was here where he taught a teenage Alexander. It seems likely that emotions were part of the curriculum. If Aristotle had had his way, any powerful young man would have learned how to manipulate them. To teach his young pupil about emotions, he likely would have begun with the entity that, according to both him and Plato, underpins and generates our feelings—the soul.

On the face of it, Aristotle's and Plato's theories of the soul are very similar. Both saw the soul as existing in three parts, and both thought each part had a different set of powers. Both thought reason was located in a particular section of the soul that only existed in humans and gods. But Plato believed that a soul was external and controlled the shadowy body. Aristotle didn't.

Aristotle's three parts of the soul were based on what he observed. He saw that plants, animals, and humans were all alive, so he figured that all must have some sort of soul or life force. He saw that plants reproduce and grow as humans and animals do, but they don't "feel" emotions or move very much. The soul of a plant must be simple, he reasoned, a "vegetative" soul. He also observed that animals do all the things plants do, but they also feel sensations and move, as humans do. They therefore must have a sensitive part of the soul. What animals can't do is write philosophy and think about what souls are—they just react. He figured they must be lacking the rational part of the soul, which does the thinking. These three parts of the soul work as a hierarchy. Plants only have vegetative souls. If you add a sensitive soul to that, you get an animal. Finally, if you add a rational soul to an animal, you find yourself with a human.

Aristotle didn't think that these souls were the real you and that your body was just some shadowy reflection on a cave wall, as Plato did. He thought that a body without a soul was just a lump of stuff—lifeless flesh that will soon become dirt. Similarly, he said, the soul has to be "*in* a body."[20] A soul without a body would cease to exist. The body and the soul create life together. The more parts of a soul something has, the more complicated a form of life it is. If it can reproduce and grow, it has been shaped by a vegetative soul. If it can also feel, sense things, and react, it's shaped further by a sensitive soul. If it can think, it's shaped further still by a rational soul.[21] Similarly, a soul can't exist without a body. Life needs both.

Aristotle didn't agree with Plato's idea that all parts of the soul could feel. For Aristotle, only the sensitive part could be perturbed and feel *pathē*, because that's the part that controls the things that cause those disturbances—the external senses of sight, smell, touch, taste, and hearing. Aristotle thought it worked like this: your senses pick up on something in the outside world; maybe you see a lion. Seeing the danger perturbs your sensitive soul, causing *pathē*, perhaps of fear. Then you might flee immediately, with no input from your reasonable soul. Or your reasonable soul might step in to remind you that you're at a zoo, so the lion can't harm you. This pattern is the same with all the senses, which act as a link between the outside world and your internal feelings.[22]

One of the most significant differences between Plato and Aristotle is that Aristotle didn't think emotions needed to be suppressed for the greater good—at least not in the lofty, almost spiritual, way that Plato did. If you were going to control your emotions, Aristotle thought, it should be to use them in a debate style called rhetoric. Rhetoric is the art of persuasion, and at its core is something called *pathos*, or the ability to draw out the feelings of the people to whom you are speaking. It's a type of argumentation that politicians and lawyers still rely on today, perhaps too often.

Rhetoric doesn't mean ignoring the facts altogether. In fact, in Aristotle's book (or, rather, notes) on the subject—unsurprisingly titled *Rhetoric*—he repeatedly enjoins his readers to know all they can about the topics they intend to debate. But the facts should be presented using a range of emotional tricks, developed to win over your interlocutors emotionally and intellectually.

What makes *Rhetoric* useful for us is that Aristotle spends a fair amount of time detailing certain emotions. (Plato does this as well, but in a less systematic way.) Specifically, these are the *pathē* useful for *pathos*. He discusses them as pairs of opposites in book 2 of *Rhetoric*. But before he gets to them, he ends book 1 with the two primary emotions that he thinks all the others come from.

- The first is pleasure, which he defines as "a movement by which the soul as a whole is consciously brought into its normal state of being"—in other words, an unperturbed soul. Pleasure is associated with those feelings that help the soul go back to being unperturbed.[23]
- The second is pain—the opposite of pleasure, or a perturbed soul. Pain is associated with the feelings that cause the ripples in the first place.[24]

The ability to manipulate the pleasure and pain of others is what makes a good rhetorician. In some ways, this is what Socrates was doing. If you read Plato's dialogues, you'll find that Socrates's endless interrogations provoked intense, painful reactions—most often anger. Perhaps that's why anger is the first discrete *pathē* Aristotle wrote about.

Aristotle's list of emotional pairs useful for rhetoric is:

- anger/calm;[25]
- friendship/enmity (or love/hate);[26]
- fear/confidence (the latter is described as "the absence or remoteness of what is terrible: it may be due either to the near presence of what inspires confidence or to the absence of what causes alarm");[27]
- shame/shamelessness;[28]
- kindness/unkindness;*[29]
- pity/indignation;[30] and
- envy/emulation (the latter is described as a sort of good envy— feeling pleased when somebody has gotten or achieved something).[31]

According to Aristotle, the ability to express these feelings, and manipulate them in your opponent, goes a long way toward helping you win an argument. You had to master them, both in yourself and in others, if you wanted to be a debater with the skill of Socrates. Just be careful whom you debate with; it could get you into trouble.

Rhetoric to Alexander

Knowing that Alexander had studied with Aristotle and absorbed his views on emotions and rhetoric gives us insight into how he likely

---

* This is interesting because it suggests that Aristotle thought of a person's actions, not just feelings, as the result of *pathē*.

thought about his letter from Darius. Darius was furious at the notion of Alexander trying to invade his lands. Alexander was well aware that any response to this letter other than "Okay, I'm leaving" would get him into trouble. Or at least he hoped it would. Trouble appears to be what he was after. That and an empire the likes of which the world had never seen. He was going to get both.

Alexander is said to have read the mocking, insulting letter from Darius to his soldiers, which apparently worried some of them. After all, the Greeks had a long, brutal history of war with the Persians. But Alexander drew on his old teacher's wisdom and spoke to his troops: "Why have you been frightened by these words?...The chieftain, Darius, not being able to do anything by his actions, pretends to be someone in his writings, just as dogs do in their barking."[32] Darius was trying to frighten Alexander and his troops. The problem for Darius was that Alexander knew what he and his army were capable of. Aristotle might have taught his old pupil that "when we are convinced that we excel in the qualities for which we are jeered at, we can ignore the jeering."[33] "The absence of what causes alarm" was Aristotle's definition of confidence, and Alexander exuded it.

Understanding your audience is also essential to rhetoric. Aristotle claimed that you "must also take into account the nature of [your] particular audience when making a speech of praise; for, as Socrates used to say, 'it is not difficult to praise the Athenians to an Athenian audience.'"[34] Perhaps with that in mind, Alexander pointed out to his troops that the letter contained evidence of Darius's immense wealth, so his men could look forward to sharing the loot when they beat him. The gathered combatants rather liked that idea. Alexander had replaced his troops' fear with the same confidence and desire he was feeling, just as he had been trained to do.

Once his men were calmed, there was nothing for Alexander to do but write back.

It is the shame of shames that so great a king as Darius, who is bolstered by such great forces and shares the throne of the gods, [will] fall under the humble and abject servitude of a single man, Alexander.[35]

Alexander then told his men to crucify the Persian messengers, who, quite understandably, were "panic-stricken." He was never going to actually have them killed, however. After some groveling on their part, Alexander released them and told them:

Now you are terrified and afraid of being beaten to death and beg not to die; so I am setting you free. For my desire is not to slay you but rather to show the difference between the king of the Greeks and that barbarian tyrant of yours.[36]

Darius had made a mistake; Alexander had been taught well. He knew how to manipulate people's feelings. He could inspire bravery in his men by mocking Darius's attempts at intimidation. What's more, he gave Darius a lesson on the proper way to terrify your enemies: playing with a man's desire to stay alive before releasing him from mortal peril has great power. The messengers went back to tell Darius that Alexander was great.

Almost ten years later, Alexander stood by the Ganges River, in northern India. He had conquered Egypt and much of the land between Greece and the Himalayas, including Darius's kingdom, just as the gods had promised. But his troops had finally had enough. They missed their families and their homes and refused to battle any further. They wanted to go home. Alexander dreamed of what might lay beyond, what other treasures he could behold. But there, by the Ganges, he reluctantly agreed to set off back to Greece. His empire had reached its limit.

It wasn't so bad going home. Alexander could catch up with his old tutor, Aristotle, and read some of the dialogues that he'd been writing at his new school, the Lyceum.

Feeling Philosophical

Plato and Aristotle are two of the biggest names in ancient Greek philosophy. Emotions were of vital importance to both of them, although, as with so many other things, they disagreed not only about how they worked but also about how they ought to be used. To Plato, feelings could either raise us to a greater good or condemn us to dangerous short-term pleasures. Aristotle thought emotions sprang from a part of the soul we share with animals and that they were useful when arguing or negotiating with an enemy. Both believed that emotions could be manipulated by reason, or *lógos*. Plato thought emotions should be directed toward something high, something spiritual, while Aristotle thought of them in a practical, down-to-earth way, as a tool for getting things done. Pretty much every disagreement between these two men can be boiled down to Plato's emphasis on the spiritual versus Aristotle's pragmatic focus on real-world applications. Their views on emotion are certainly no different.

Plato's and Aristotle's ideas about emotion and the soul came to form the cornerstone of Western thought and politics for nearly two thousand years. One or the other, or both, influenced every philosopher who followed, as well as civilizations, political movements, and religious beliefs. The theories of emotion that Plato and Aristotle offered helped to establish the cultures and beliefs of the entire Western world. Don't believe me? Keep reading, and you'll see how often their basic ideas surface and resurface. They helped millions of people around the world understand themselves, and it wasn't until the 1600s that anyone seriously challenged them.

Needless to say, the ancient Greeks weren't the only people who influenced theories about emotion. Nor were Plato and Aristotle the only people to have a significant global impact by trying to spell out what emotions are and what we should do about them. In ancient India, at roughly the same time that Socrates was facing his execution, another debate about feelings was taking place, one that also appears to have inspired just about every thinker who came after it, from the shores of China to the fringes of Christendom. So let's move to the same northern provinces of India from which Alexander turned back—the place where an emperor's emotional conversion had a profound effect on history.

Two

Indian Desires

King Chandashoka was a cruel, violent man; his brutality was infamous across all of India. According to some legends, he was so evil that he visited hell to learn the most vicious ways to torture people. Chandashoka inherited his modest kingdom—the Mauryan Empire—circa 265 BCE and immediately set about expanding the borders of his reign. Before long, his empire stretched across the Indian subcontinent from what is now Afghanistan to what is now Bangladesh. But after eight years there was still one annoying little region on the East Indian coast that he hadn't quite found a way to subdue: Kalinga.

When Chandashoka marched on Kalinga, events didn't go as planned. The war was long and brutal. Death, mutilation, and carnage surrounded him in every direction; the ground ran red with blood. His earlier battles had killed thousands, sometimes tens of thousands. This time, more than one hundred thousand people were slaughtered, and many more died of disease and starvation because of the fighting. Another one hundred and fifty thousand people were displaced from their homes. Chandashoka witnessed a train of desperate, terrified refugees forced to walk for hundreds of miles. During battle and afterward, he was exposed to evils that made hell look like paradise. And if the legends are only partly true about what happened next, this experience caused him to undergo one of the most profound spiritual conversions in history. For this reason, the world doesn't remember him as Chandashoka (Ashoka the Monster, or Ashoka the Cruel). In India, he became known as King Piyadasi (he who regards everyone with affection) and Devanampriya (beloved of the gods). He's more commonly known as Ashoka the Great.

Ashoka's conversion took place because the focus of his *desires* changed, and that was no small thing. Ideas about desire and what you should do with it lie at the root of most ancient Indian religions, but each understands desire somewhat differently. Any change in Ashoka's wants and desires was likely to dramatically affect the way he lived his life and what he believed. But as feelings go, desire is unusual and complicated. So before we move on, let me try to unpack some modern ideas about desire that might help us better understand the forces that changed Ashoka's life and, with it, the lives of his subjects.

Rewarding Desires

Imagine yourself on a beach, the sun warming your skin, a cool breeze regulating the temperature, the waves breaking gently onto the shore. You lie back, feeling the sand cascading between your toes as the stresses of daily life fall away. You feel calm; you feel happy. But as you gaze out to sea, you notice the fin of a great white shark. The calmness and happiness give way to fear and panic as you alone seem to have spotted the danger. You feel a deep, visceral desire to tell someone. You leave your idyllic spot and run for the lifeguard, who sounds the alarm and clears the shoreline before any harm can be done. Then you experience relief.

Of all the feelings you experienced, did you notice the odd one out? Most of those feelings—happiness, calm, fear, panic, and relief— were caused by the world around you affecting your senses. You felt relaxed and happy because of the heat of the sun, the brush of the sand, the sound of the ocean. You felt fear and panic because your eyes detected the shark, and you felt relief when you could see that every-one was safe. But the desire you experienced was different. Even though the sight of the shark triggered the urge to tell someone, it was the desire to stop people from getting injured that sent you running to the lifeguard. You might have felt the fear and still watched, paralyzed with terror, as someone got hurt. You might have panicked and waved your arms around, not really helping the situation. There are a million things you could have done, but instead you felt a desire to tell some-one about the shark—a desire for people to be safe. In other words, while most feelings are "world-to-brain"—that is, we sense something in the world, and it creates feelings inside us—desire is "brain-to-world." We feel desire, then we do something.

There are many types of desire. It can be intrinsic—that is, a desire you have for its own sake, such as wanting ice cream or a nice car. Or maybe you want to share your love of a subject with other people. Desire can also be instrumental—a desire to do something that will lead to something else. For example, you might desire to earn money so that you can buy an ice cream cone. You might desire to buy flowers to make your wife happy, or you might even desire to write a book so that others can be as enthusiastic about the history of emotion as you are. Desires can be strong or weak, depending on how badly you want something. They can also be occurrent; that is, they can prey on your mind. Desir-ing to pass your next exam and desiring to get out of a sticky situation fall into this category. Desires can also be standing: standing desires, such as the desire to live to a ripe old age, linger in the back of your mind.

Each of these subsets of desire can also be, as philosopher Harry Frankfurt puts it, either first-order desires—the desire for an object or for an event to take place—or second-order desires, the desire to desire.[1] Your desire to tell a lifeguard about a shark would be an instru-mental first-order desire: you desire to do something that leads to something else (to tell the lifeguard so that he can get people out of the water) and for something to happen (for lives to be saved). But while desire can be explained in terms of orders and types, there is a way to understand them that I think works better. This way of looking at desire comes from philosopher Timothy Schroeder. He managed to boil all the many philosophical and scientific ideas about wanting and wishing into what he calls the three faces of desire.[2]

The first is motivational desire: the desire to get up and do things. The second is hedonic desire: the desire to experience pleasure and avoid pain. Finally, there is learning desire: the desire to learn through experience what will be good for us and what will not. Of course, these three need not exist in a vacuum. The motivational desire to pass a driving test requires a hedonic desire for the pleasure that driving independently gives you and the learning desire to become an experienced motorist.

Schroeder also identifies two unifying elements of desire: reward and punishment. Learning to ski is a reward. Not learning to ski and being laughed at (or breaking your neck) is a punishment. Reward and punishment are what link desire not only to thoughts but also to other emotions, including some of the most primitive we have. I'll primarily be using Schroeder's framework to unpack the various types of desire from this point on.

By the way, when I use the word *desire*, I also mean things that are found under that word in a thesaurus—need, want, appetite, longing, craving, yearning, wish, aspiring, hankering, and so on—partly because it stops me from unnecessarily overcomplicating things but mostly because people throughout history didn't often differentiate between their desires and needs, either. They did have many subtle types of desire, but a modern thesaurus can't help you with those. Let me explain.

Hindu Desires

Ashoka was raised in the dominant Hindu culture of northern India. It's important to note that *Hindu* is a somewhat contested term. In this book, I use the words *Hindu* and *Hinduism* advisedly. For a start, the people who follow what we now recognize as Hinduism didn't call themselves Hindus until the British showed up in the seventeenth century. More confusingly, there is no single Hindu religion, and there never has been. One historian of Hindu religion—Professor June McDaniel—thinks there are several different Hinduisms and counts six main varieties. Another Sanskrit expert and Indologist, Professor Wendy Doniger, writes that each adherent of the Hindu faith, or faiths, "holds a toolbox of different beliefs more or less simultaneously, drawing upon one on one occasion, another on another."[3] This wonderful variety of beliefs notwithstanding, some ideas are loosely shared across all strains of Hinduism. One of these is an understanding of the types of desire.

In many Hindu scriptures there are four types of desire, or aims (*purusarthas*). The most important of these is each person's path, or *dharma. Dharma* is both a motivational and a learning desire. Its role is to keep people on the straight and narrow, allow them to grow and learn, and act as a brake on any hedonic desires they might feel. One of the best examples of *dharma* comes from the Bhagavad Gita, or song of God, which is part of the Mahabharata, a work depicting a war in the Kurukshetra region—a religious war that may or may not have happened between two powerful families.[4]

The Bhagavad Gita begins when a prince from one side, Prince Arjuna, is facing the final battle. He asks his charioteer to move closer to the enemy so he can get a proper look, and he's horrified to see members of his own family in the opposing army. Arjuna doesn't want to kill his kin, so he wonders if he can get out of the battle by evoking the divine virtue of nonviolence (*ahimsa*). The excellent news for Arjuna is that his charioteer, on top of being fantastic at his job, alarmingly handsome, spectacularly intelligent, and faultlessly kind, just so happens to be Krishna, the living embodiment of God. As you might imagine, God makes a great adviser.

Krishna explains to Arjuna that nonviolence is good, really good, but it's even better to follow your *dharma*, your true path. To follow your path is to be selfless. You have to ignore the *bhavas* that try to tempt you away from it. *Bhavas* are complicated. The word *bhava* can mean both "existing" and "being created"—life and birth. It can also refer to a mood, feeling, or habit. And it can mean all these things at the same time. The idea is that we exist because we feel. We feel because we are born. To be alive is to have feelings—otherwise you are just empty, or dead. But for you to remain true to your *dharma*, all *bhavas* have to be kept in check. You must follow your path regardless of how you feel about it. To do that, you have to renounce all the pains and gains created by it.[5] Your path is not about filling your boots with loot. It's not about stopping a fight because you like some of your opponents. It's about doing what you were born to do. What you must do. Not fighting this battle would be just as contrary to Arjuna's *dharma* as fighting only for glory and power would be. Some forms of Hinduism claim that the reward for keeping to your *dharma* is reincarnation into a better life or reaching a moment of bliss in this one. The punishment for not following your *dharma* is the opposite. Arjuna had to put all his feelings to one side and do his duty.

Although *dharma* is the most important desire to follow, it's not the only one. The next is *artha*. This is a motivational desire for the things you need to live your life. It's about the acquisition of wealth, a home, and all the other bits and pieces that matter in day-to-day living. The precise meaning of *artha* can also be a bit contradictory and difficult to pin down within the Hindu scriptures. And *artha* can change depending on a person's need. For example, if you're religious, your idea of wealth might be devotion to the gods. If you're a politician, it might be the consolidation of power.[6]

One work discussing this desire, the *Arthashastra*, was written around 400 BCE and is primarily a pretty brutal treatise about politics. It teaches rulers how to get what they need by tricking people. I'm not just talking about the mundane stuff, such as using a technicality to claim that you spent more on health care last year than you actually did. It contains, for example, advice on consolidating your power by predicting someone's death just before that person is mysteriously executed—by you—thus proving that you can see into the future. What's important is that, to use Harry Frankfurt's definitions for a moment, your first-order *artha* should never conflict with your second-order *dharma*. *Artha* is a desire for things; *dharma* is the desire to follow your path, regardless of things. But if being a king who has to be brutal in order to keep order and power is part of your *dharma*, then an *artha* that allows you to build torture chambers and armies is fine.[7]

The third desire is *kama*. *Kama* is not to be mistaken for *karma*. *Karma* is a summation of your deeds across this and previous lifetimes. It governs whether, how, and into what you will be reincarnated. *Kama*, however, is a hedonic desire for worldly pleasures. You might know *kama* from the *Kama Sutra*, the famous text that is not as much about sex as you think it is. It holds a great deal of wisdom about how best to control and fulfill desires, though it also has a lot to say on the subject of coitus. *Kama* is not just a garden-variety lust for something. It's a force, a power that pushes all living things into making decisions and choosing paths. We overeat because of it, sleep too much because of it, indulge in sex and drugs and rock 'n' roll because of it. The rewards for satisfying this type of desire are immediate, as are the punishments. But pursuing *kama* at the expense of your *dharma* leads to punishment in the next life. Once again, it's a choice between first-order and second-order desires.

Finally, *moksha* is the desire to know your true self. To explain this, I need to explain something about the way Hindus understand the soul.

Most Hindus believe that we are made up of our true selves—the *atman* and its surrounding five sheaths, or *kosha*. The *atman* is part of Brahman—a Hindu word for God. Put simply, to know your true self is to know that bit of you that is also part of God, or that bit of yourself that's godlike. This might sound a bit like *lógos*, but whereas the *lógos* is accessed through thinking, the *atman* is harder to reach because it lies beyond thought. To get to the *atman* you have to work your way through the *kosha*, comprising the following layers.

- The food sheath (*annamaya kosha*) is the stuff of our bodies, made from the food we eat. This is the outermost sheath.
- The vital sheath (*pranamaya kosha*) is the "glue," the air that holds the sheaths together. The control of air was and is especially vital in the yogic tradition, and that's why your yoga classes still begin with lots of slow breathing.
- The mind sheath (*manomaya kosha*) is why I am "I" to me and why you are "I" to you. It's the sheath that makes us who we are.
- The intellect sheath (*vijnanamaya kosha*) is the thinking bit. With it, you can make smart decisions, such as buying this book. This is where you find thought—where you are closest to the *atman* without being in direct contact with it. It's the closest thing to the Greek notion of *lógos*.
- Finally, the bliss sheath (*anandamaya kosha*) is a reflection of your *atman*. This sheath is the most important, because by tapping into it, you can achieve bliss and real happiness (*sukha*). If you achieve that, your next life will be better. Some people even think you can permanently avoid reincarnation.

Moksha, then, is a motivational desire to find a path through these sheaths and rediscover your *atman*. There are a few ways to do that. One way is to eat as little as possible, wear simple clothes, move to the forest, and meditate twenty-four hours a day. If you don't fancy stripping almost naked and starving yourself half to death in a forest right away, you could use it as a retirement plan. Simply follow your path until you're older and your *dharma* reaches its near conclusion, then do the *moksha* stuff. Alternatively, you could support people who strip and live in the woods and allow them to discover their *atmans* on your behalf. Another way to renounce is by simply following your *dharma*. The reward is to know your true self, which is the goal of *dharma* and another of its rewards.

These four desires—*dharma*, *artha*, *kama*, and *moksha*—are important because they can take you in one of two directions. The first involves using one of the methods described above to find genuine bliss through the knowledge of your true self. The other way might appear to lead to joy, but it leads to its opposite—*dukkha*, or suffering. The second path is easy. You let your *kama* take charge and guide you toward the pleasures of the body. You lust for sex, good food, and good times. Or you let your *artha* lead, striving to become rich and buy a big house. But by doing this, you neglect your *dharma*, and you don't renounce anything—unless, of course, your *dharma* is to be a party animal or a rich man,* which is possible but unlikely. The problem is, unless you are lucky enough to be destined to be a party animal, you can never feel bliss this way. Even if you get everything you want, greed (*lobha*), anger (*krodha*), envy (*matsarya*), and fear (*bhaya*) will plague you every step of the way. Always wanting more and fearing the loss of what you have will lead you, ultimately, to suffering in this life or the next. To achieve real bliss, you have to balance those desires, follow your *dharma*, and fulfill *moksha*.

It could be argued that Ashoka the Monster was originally trying to follow his *dharma*. Although they might be missing the point of what Krishna was trying to say, outside observers might think that Ashoka was following his path as laid out from his birth. Being born into a conquest-focused royal family meant that expansion and invasion were his *dharma*. Through *artha*, he maintained his power and he gained those material things he needed to be a king—palaces, armies,

---

* And by the way, it is always a man. Women at the time didn't follow any *dharmas* except that of being a good wife.

torture chambers. He believed they were necessary for his empire's continued control of the region. He was also likely to have indulged in a little *kama* from time to time with his five wives, as his *dharma* dictated.

Viewed in this way, Ashoka's focus on his *dharma* somewhat parallels Arjuna's story. And yet, surrounded by the bodies of the Kalingan dead, he seems to have fallen into the most profound sense of sorrow. No matter how many lives he took, how many wives he had, or how much land he conquered, his desires weren't leading him to any sort of bliss. All he'd ever known was pain and suffering. Perhaps he misunderstood or didn't care about the core point of this type of Hinduism—that your *dharma* is bigger than your own feelings in the moment. Whatever the reason, he chose another path. He turned to the Buddha.

The Boy from Lumbini

Around two hundred years before Ashoka converted to Buddhism, what was left of a man named Siddhartha Gautama stumbled down the banks of the Phalgu River near the city of Gaya, in the northeast of India. His muscles were withered, and his skin stretched over his bones. He looked like a skeleton wearing a tight leather overcoat. His eyes were sunk back in his head, his hair was thinning, and he was tired—so very tired. But although his body was wrecked, he was far from broken. In fact, he'd never felt so good.

Gautama's young life was sheltered, and he was rarely allowed out of the palace in Lumbini, where he was raised. He was lucky enough to be born a *kshatriya*—a prince of the warrior and kingly class. The only other castes of people he met were the priests—the *brahmins*—and occasionally merchants and landowners, the *vaishya*. He longed to know how the peasants—the *shudra*—and those without a caste, the untouchables, lived. When he was twenty-nine years old, he sneaked out of his palace with the aid of his charioteer to see what was going on beyond the walls that had trapped him for so long.

He soon discovered what a privileged life he had. First, he saw an old man. Gautama was shocked at the devastation and suffering caused by the ravages of time. Next, he saw a sick person suffering in the pain of his illness. Then he came across a dead body. He was repulsed and horrified by both. Those encounters shook him to his core. The dreadful cycle of life and death that he and everyone else was trapped in was almost too terrible for him to bear. After he returned home, he knew he had to find an escape, a way to move beyond that desperate fate. So he sneaked out of the palace again. This time he met an ascetic—a man who had devoted his life to trying to find a way out of the suffering. On this encounter, Gautama realized he wasn't alone, and he felt hope. He resolved to join the ascetic on his quest.

Gautama left the palace, his wife, and his children behind and spent years searching for a way to end suffering and break the cycle of death and rebirth. He tried one method of yogic meditation after another, but nothing worked. Then he remembered an experience he had as a child. One beautiful summer day, he sat in a yogic position, watching a plowing competition under a rose apple tree. He noticed the cut grass and the dead insects and felt horror at the violence of it all. Then something about what he saw brought release. It was as if the stillness he witnessed in the aftermath of that violence helped him see beyond the cycle of life and death. In the face of that destruction, the young Gautama knew that life, death, suffering, and the world surrounding him were illusory. He melted away from the illusion into a moment of pure bliss. When he remembered that childhood joy, he stood up, ate a bowl of rice pudding, and set off to find a tree to sit beneath. By the banks of the Phalgu he found a sacred fig tree, and he sat and meditated just as he had when he was a child. After three days and nights under that tree, Gautama achieved that same childhood state of enlightenment: *nirvana*. He had been released from the cycle of life and death. He became the Buddha.

Or at least that's the legend. Gautama may or may not have been a prince, and he may or may not have been sent on that path after seeing an old man, a sick man, and a dead man. We don't know for sure. We do know that he spent time as an ascetic, and he probably did remember a moment of joy from his childhood. But he almost certainly didn't set off for the tree, now known as the Bodhi Tree (or "tree of enlightenment"), and work it all out in just a few days. It would have taken weeks, even months, for his body to recover from the ravages of the starvation diet he'd been following. It's more probable that he spent some time thinking and developing his ideas before finally hitting on the big one.

Most of his new idea wasn't really all that new. What the Buddha did was to draw on the beliefs of his culture and reinterpret them, distill them. What he came up with is now known as the Four Noble Truths, the first three of which we've already met.

1. There is suffering.
2. The cause of suffering is desire.
3. The way out of suffering is through *nirvana* (or, in Hinduism, connecting to your bliss sheath).[8]

So far so good. But the last noble truth was something unique: there is a path to *nirvana*. This path came to be known as the Noble Eightfold Path.

Sometimes the Eightfold Path is split into three divisions. The first division is basically about becoming a Buddhist, and the second is about being a good person. The third division is the one that gets you to *nirvana*, and it's all about desire. Or at least something like it. To explain, let's visit a legend from later in the Buddha's life.

Ancient Buddhist Emotions

When the Buddha was quite old,[9] he went to live in a town in northern India called Vaishali.[10] This area was known for another faith—Jainism, a religion with a deep focus on asceticism and nonviolence toward all living beings. One of these Jains, Saccaka, claimed that he could beat anyone in a debate. He boasted that even a "senseless post" would "shake, shiver, and tremble" with the force of his arguments.[11] On hearing that the Buddha was in town, he naturally challenged him to a debate, threatening to intellectually "thump him about."[12]

Not a shy man, Saccaka found an audience and led them to the hall where the Buddha was staying, ready to do battle. The Buddha answered Saccaka's opening question. Then he asked Saccaka one of his own: "Is it possible to have control over your body, senses, feelings, thoughts, and consciousness?" Saccaka fell silent. The Buddha asked again. Nothing. The Buddha asked one more time. Still, Saccaka had no answer. The Buddha took this as a teaching moment:

> It was you who made this statement before the Vesīlī assembly...Now there are drops of sweat on your forehead and they have soaked through your upper robe and fallen to the ground. But there is no sweat on my body now.

Saccaka "sat silent, dismayed, with shoulders drooping, and head down, glum, and without response." Why did the Buddha remain calm while Saccaka became sweaty and drooping? And what is the relevance of the five things the Buddha asked him about—body, senses, feelings, thoughts, and consciousness?[13]

The Buddha did not sweat because he had mastered the seventh step on the Eightfold Path: right mindfulness, or being mindful. This step is about being aware of everything going on around you, and it isn't far from the practice of mindfulness that many mental health experts prescribe now. To be mindful is to pay attention to everything in your surroundings. It's about not dwelling on the past or dreaming of the future; it's about being in the "moment." Ancient Buddhist practitioners of mindfulness achieved this through a mastery over what they called the Five Aggregates of Clinging, or *skandhas*. This is the answer to the Buddha's question—how to control your body, senses, feelings, thoughts, and consciousness. At the nucleus of all these things are two types of desire.

The first of these desires is clinging (*upadana*). Clinging is a sticky sort of hedonic desire that, according to traditional Buddhism, consumes us at every level. It's why we hold on to existence, getting continually reincarnated rather than reaching *nirvana*. *Upadana* also makes you stay silent rather than admit you can't answer a question and ruin your reputation. Like the Hindu concept of *kama*, clinging can never be satisfied. It will always be seeking "this and that, here and there (*tatra tatra abhinandini*)."[14] Clinging is made up of a few parts, or aggregates.

The first aggregate of clinging is your body, or form: *rupa*. The body itself can't be controlled, but it's thought to be a window on your other aggregates. In the foundational Buddhist texts, there are descriptions of "lean, wretched, unsightly" people who are "jaundiced, with veins standing out on their limbs," signifying that they're up to no good.[15] By keeping mindful of your body—noticing if your veins are standing out or you're starting to sweat—you can track the other aggregates: *sanna*, *vedana*, *vinnana*, and *sankhara*.

For your body to react, there must be something for it to react to. At the start of this process is perception, or *sanna*. This is the ability to understand the world around us through our senses. Such perceptions cause undesirable feelings, or *vedana*. *Vedana* are basic sensations that produce pleasure, pain, or those neutral feelings that aren't all that painful, but nor are they particularly pleasant—bland feelings, like the experience of watching paint dry. If you don't control the pleasurable sensations, they lead to darker hedonic desires such as lust, greed, and obsession. If you don't restrain your painful feelings, these can turn into anger, fear, and sorrow. If you fail to master your neutral feelings, you'll get bored. The Buddha never lost control of his feelings because, through the application of a second-order motivational and learning desire we'll get to in a moment, he knew how to keep them in check.

The next aggregate is consciousness, or *vinnana*. It works closely with the other aggregates, but they can lead it astray by letting in three poisons — *moha* (delusion, confusion), *raga* (greed, attachment), and *dvesha* (aversion, hatred). During Saccaka's argument with the Buddha, his delusion made him experience painful feelings, sweat profusely, and hang his head. The problem is that these first four aggregates are beyond your direct control — unless you learn to master the second type of Buddhist desire, *chanda*.

Chanda is central to Buddhism. It's a desire that helps you follow the Eightfold Path by keeping clinging at bay. To put it another way, in a strange twist on second-order desires, *chanda* is a desire not to desire. The eagle-eyed among you might have noticed that *chanda* was the word tagged onto Ashoka when he was known as Chandashoka — Ashoka the Monster, or Ashoka the Cruel. That isn't a coincidence. To the Buddha, *chanda* was a monster. It was cruel, torturous, and brutal. No one ever said attaining *nirvana* was easy. But taming this beast was essential if you wanted to control your thoughts, or *sankhara*.

Sankhara is the cog around which the other aggregates of clinging spin. It's also the foundation of *chanda*. If thoughts are left unchecked, they can spin out of control. Saccaka wasn't able to control his *sankhara*. He hadn't mastered *chanda*, so he had no desire to control his thoughts and feelings. He felt a craving (*tanha*) to prove how clever he was. Craving causes lust, greed, and obsession and can lead to clinging, which in turn can lead to *dukkha* (suffering).

The good news is that Buddha taught everyone how to master

chanda. You begin by sticking to step number 6 in the Eightfold Path—right effort. Replace greed with gladness at others' success (*mudita*). Throw away hate and restore loving-kindness (*metta*) and compassion (*karuna*). Then move to the seventh step, right mindfulness. Move away from the delusion of existence and surround yourself with the Buddhist insight (*punna, vipassana*) that—in contrast to Hindu beliefs—there is no true self. In fact, there's no self at all. Once you've managed that, you are ready to take the final step—right *samadhi*, a particularly deep form of meditation that can lead to *nirvana*.

The Buddha realized that people were consumed by the wrong sort of desire. What kept people in the cycle of life and death, he believed, was clinging and the aggregates that cause it: your body (*rupa*), your understanding of the world formed through your senses (*sanna*), your undesirable feelings (*vedana*), and your unchecked thoughts (*sankhara*). These kept the focus on this world, a world that, the Buddha believed, was an illusion. To escape the illusion, you had to focus on the right sort of desire—*chanda*—and use it to follow the Eightfold Path. The Buddha remained calm during the debate because his *chanda* was perfect and focused; Saccaka's wasn't. By desiring to control his mind, the Buddha held his feelings down, and his body followed. The reward would be enlightenment, *nirvana*—the chance to escape the cycle of life and death.

Ashoka's Behavior After His Conversion

Ashoka reached a turning point in his life at Kalinga. The types of desire he was raised with—those of the Hindus—weren't working for him. Avoiding sorrow by sticking to his path only seemed to cause more sorrow. He clung to his *dharma*, but it didn't lead to bliss. When Buddhist missionaries explained to him that clinging was the problem, the notion had to have been appealing. Ashoka's transformation as he went from one understanding of desire to another was quite dramatic.

The reason we know what we do about this transformation is that its history was written in stone. I mean literally written in stone. Ashoka had a series of thirty-three edicts carved into rock and stone pillars in more than thirty sites across his empire. Some were in the language of the locals. Some were in his own language. Some were carved in his neighbors' languages, such as ancient Greek and Aramaic. One of these stone edicts tells us what happened during the war to change him: "On conquering Kalinga, the Beloved of the Gods [Devanampriya] felt remorse, for when an independent country is conquered, the slaughter, death, and deportation of the people are extremely grievous to the Beloved of the Gods."[16] The campaign against Kalinga had shaken Ashoka to his core and made him reevaluate his life and his religion.

Ashoka himself said that he had been working on becoming a Buddhist for a little while before he took it seriously. One of his edicts reads: "I have been a Buddhist layman [upasaka] for more than two and a half years, but for a year I did not make much progress. I have drawn close to the community [of monks, or samgha] and have become more ardent."[17] His commitment to his new faith manifested itself not only in declarations carved in stone but also in real, practical support. He built Buddhist monuments (stupas) and temples across his land. He sent educators to teach Buddhism to people in his kingdom who couldn't read his edicts and missionaries to his neighbors to spread the faith. He stopped eating meat and made a long list of the types of animals his subjects were and were not allowed to eat. He replaced the violence of the traditional royal hunt with the pacifism of a royal tour. He planted mango groves and sunk wells near roads for weary travelers. He made a pilgrimage to the Bodhi Tree itself. He may even have hosted the Third Buddhist Council, where infiltrators of Buddhism from other religions, determined to discredit their emperor's new faith, were weeded out and the best bits of the Pali Canon, the great ancient Buddhist scriptures, were woven into the texts. Ashoka was a changed man.

The new desire that had become central to Ashoka's life probably had an impact beyond his empire. Some of the areas the missionaries are supposed to have visited—such as modern-day Sri Lanka, Tibet, Myanmar, and Thailand—are, to this day, primarily Buddhist. Some historians, such as Professor Romila Thapar, professor emerita at Jawaharlal Nehru University, think the spread was more attributable to the traders, merchants, and guilds who supported the spread of Buddhism into these areas rather than Ashoka's missionaries.[18] But whether he had a small impact or a large one, he was almost undoubtedly influential. A moment of suffering in the thick of battle and a change in his understanding of desire are what prompted Ashoka to supercharge what was up until then a fairly minor belief system. Emotions, not actions, underlay the rise and the doctrinal development of one of the world's great religions.

An Emotional Path to Christianity

Despite their superficial differences, there are many similarities between the ancient Greek and ancient Indian conceptions of emotion. Both see pleasure and pain as crucial. Both say that desire can be dangerous, and both believe that it must be controlled. Both suggest that thoughts and feelings are usually intertwined, as are actions and perceptions of the world. Of course, both these ancient views of emotion are quite different from the way we understand emotion today.

Buddhist understandings of emotion shaped the world in the East, where the understanding of desire that grounds Buddhism would travel throughout Asia, spreading into China to become part of the emotional bedrock of a unified nation.

In 2021, there were around 535 million Buddhists worldwide.[19]

These Buddhists' belief system is rooted in the Buddha's understanding of emotion and his desire to get rid of all desire — except for the desire to not desire, of course. It was the effect feelings had on the Buddha that eventually helped shape much of Asian history. His concepts of emotional control, even now, buttress most Asian ideologies. We'll see in later chapters how they still influence Japanese and Chinese thought nearly two thousand years after their formation.

The same is true of many types of Hinduism, only more so. In 2021, there were around 1.2 billion people worldwide struggling to remain true to their *dharmas*, a struggle that's an emotional one. Hinduism, in all its beautiful shades of complexity, is a religion of desires. Like Buddhism, it's about the right sort of desire. Not necessarily the desire not to desire, which belongs to Buddhism, but the desire to fulfill your *dharma* — to be what and who you should be. And an entire subcontinent of people and beyond live and work and vote with that in mind. But that sort of emotional control isn't unique to religions of the East.

These Indian ideas very likely gave us the Greek philosophy called Stoicism. It was this that Saint Paul mixed with his own people's (the ancient Hebrews') understanding of emotion to create a strand of Christian thought that remains dominant to this day. So let's walk in the footsteps of Saint Paul as he adapts Jewish ideas about feelings for Roman sensibilities to help kick-start another of the world's great religions.

Three

The Pauline Passions

In roughly 58 CE, a man called Paul entered the Temple in Jerusalem. According to an apocryphal book called *The Acts of Paul and Thecla*, he was small, with unsteady legs, bushy black eyebrows, a thick dark beard, and an oval face set with piercing, almost otherworldly, eyes. It was a look that gave him great presence. His striking appearance was accompanied by a fierce directness and almost unshakable self-confidence.[1] He liked to talk, and people tended to listen when he spoke.

Everyone at the Temple that day had heard of him. They'd been told he was a troublemaker—that he was not only converting Gentiles to his new version of Judaism but was also telling them that they didn't need to follow Jewish customs or get circumcised. To them, this meant that he was disobeying Mosaic law and had turned his back on Judaism. Paul wanted to prove to them that nothing could be further from the truth, so he shaved his head and entered the Temple for a seven-day purification ritual. He wanted to show the Jews of Jerusalem that deep down he was still one of them.

It backfired spectacularly. Toward the end of the week, word got around about who the new guy was. As the identity of the Greek-speaking stranger spread through the Temple, the room exploded in rage. Someone in the crowd shoved him with quite a bit of force, shouting for help.

Fellow Israelites, help! This is the man who teaches everyone everywhere against our people, our law, and this place. What's more, he also brought Greeks into the temple and has defiled this holy place.[2]

Paul was yanked out of the building and dragged through the city streets. He had no doubt that he was heading for an impromptu execution. Luckily for him, a centurion and his guard broke through the baying crowd and arrested him.

These days, that sort of mob rage isn't often inspired by a lone recreant's actions. But the people in the Temple saw Paul as more than just a troublemaker. He was the worst thing a man could be—an idolater, a man who was openly defiling their house of worship and their traditions. Disgust (or something like it) and the need to cleanse Paul from the earth consumed them—the ancient Hebrews had an emotional regime of their own, and Paul was in violation of it. He's lucky that the Romans showed up when they did.

The attack on Paul is a vivid demonstration of the Pharisees' emotional regime, which Paul lived under—one of the two emotional regimes that governed his life. (I'll be explaining this group in a bit more detail further on.) The other was Roman. This dual identity played no small part in his importance to the history of Christianity. Paul's influence on Christianity is incalculable. For a start, his letters in the New Testament—the Pauline Epistles—are second only to the Gospels in how often they are quoted by practitioners of the Christian faith. On a more practical note, without his insistence that Christianity was a religion for all, rather than just for the Jewish people, it's unlikely it would have had the massive global impact it did.

How this son of a tent maker from what is now southeastern Turkey managed to become so important is often debated. But one aspect of his story that is often overlooked is the fact that he successfully merged the way the Hebrews understood their feelings with the way the

Greeks understood theirs. We're going to take a trip through the two emotional worlds of Saint Paul, starting with the one into which he was born—the one that tried to kill him in Jerusalem. But first, let's get to know Paul a little better.

The Life of Saint Paul

Paul—or Saul, as he was known to his fellow Jews (and traditionally in texts until he was converted)—was born somewhere between 5 BCE and 5 CE in the city of Tarsus, in the region of Cilicia. Despite being a province of Rome, Cilicia was a melting pot of ideas and culture: its population was mostly a mix of Jews and Greeks. The Greeks didn't know Paul as Saul: rather, they called him by his Romanized name, which would have been Something Something Paullus (a bit like Gaius Julius Caesar). We have no idea what Paul's "Something Something" was. But the fact that he had a Roman name at all means that his family was likely *civis Romanus*, or Roman citizens. That gave Paul certain rights, such as the freedom to travel and protection from angry mobs.

Though he lived in a Greek-speaking part of the world, Saul was raised as a zealous member of a faction of Jews called Pharisees. His parents tried to keep him away from any newfangled Greek and Roman ideas that his citizenship might attract. They homeschooled him, teaching him about the Pharisees' belief in the coming Messiah and the day of judgment. When he was older, they sent him to Jerusalem to study under a legendary teacher named Gamaliel. Paul was a model student, and he quickly became a gifted lawyer who seemed destined for a seat on the Great Sanhedrin, or Jewish supreme court. One of his first jobs was to witness, and possibly even take part in, the trial and execution of a man named Stephen. Stephen was a member of a new breakaway sect of Jews who believed that the Messiah had already come. After that encounter, Paul became more than a little obsessed with this new clique, a group we now call Christians.

Gamaliel had shown some leniency toward them, but this tolerance didn't rub off on his protégé, who became convinced that Christians and Christianity had to be stamped out.[3]

By the time Saul was in his early thirties, he devoted nearly all his time to uncovering hidden Christian sects around the area of Jerusalem. After rooting them out, his job was to use his considerable legal knowledge to persecute them. On a good day, that meant harassing them out of the area. On a better day—for him, at least—he'd put them on trial for blasphemy. On the best days, he would get them executed. It's fair to say that his exposure to Christian ideas was not small—he almost certainly knew and understood the central doctrines rather well, using his insight into what he considered evil and heretical beliefs to weed them out. Then everything changed.

One day, Saul was escorting a group of captured Christians from Jerusalem to Damascus in order to deliver them to the authorities. But before he reached the city, he was suddenly surrounded by a bright light and heard a voice asking, "Saul, Saul, why are you persecuting me?" Saul, confused but clearly aware that this was the voice of a being with great power, responded, "Who are you, Lord?" The voice said to him, "I am Jesus, the one you are persecuting...But get up and go into the city, and you will be told what you must do." The people traveling with Saul heard the voice, but they didn't see the light. For the following few days, Saul saw nothing, either, having been rendered blind by the power he'd glimpsed. Two days later, a Christian by the name of Ananias traveled to the house Saul was staying in. Saul knew he was coming. While praying, he'd received a premonition of the visit, and so he let him enter and place his hands over his eyes, at which point "something like scales fell from his eyes, and he regained his sight." He was baptized and from then on called himself Paul.[4]

There is some disagreement about the story of Saul's conversion on that road. Most of this story is found in the book of Acts, a work of unknown authenticity and authorship. But whether you believe that the story is completely true, a metaphor, a hallucination, the result of a lightning strike, or the result of an epileptic fit, it changed Paul and Christianity forever.[5]

As a Pharisee, Saul would have already believed that the Messiah was going to come, heralding the end times, when everyone will be resurrected from the dead and judged.[6] But on the road to Damascus, whether through meditation and prayer, a near-death experience, or divine intervention, he realized something: the man these Christians kept yammering on about—Yeshua ben Yosef, better known today as Jesus—fit the description of the Messiah in every way. It was a revelation that not only meant that Jesus was who they said he was but also that the end times had begun. Saul had work to do.

From that point onward, Paul became as zealous in his Christian faith as he had been in his Judaism. He used his freedom in order to travel much of the Roman Empire, preaching Christianity and bringing people into the religion. The problem was that he preached to everyone, and many Jews and Christians didn't like that. The Jews, not surprisingly, thought that Christianity was heresy, an opinion Paul shared before his conversion. But his fellow Christians also thought preaching to the Gentiles was heresy. In those early days, Christians believed themselves to be a Hebrew sect whose teachings should remain exclusively Jewish.

Christianity has never been a single monolithic belief. It fractured into factions almost as soon as reports of Jesus's resurrection began to circulate. But one thing Christians agreed on was that the only real difference between them and other Jews is that Christians believed the Messiah had come. Other Jewish people thought he was still on his way. If the Messiah had come, as Paul fervently believed, then the end times would soon follow. People had to be told: they needed to be saved, and to Paul, that meant everyone. Paul began to spread the word to Jew and Gentile alike, which required tapping into people's emotions. It was a risky business.

Ungodly Aversions

As I said above, Paul grew up under the yoke of two competing emotional regimes. The first of these came from his Jewish heritage. What makes this regime interesting is that we have some idea how Jewish people in this period actually *felt*. You'd be surprised by how rare this is. Most historical sources that describe feelings only talk about what emotions *do* or what people think they *are*. Ancient Greek and Indian texts are a case in point. We can guess that the emotions associated with *eros* ultimately make you feel better than those associated with *boulesis*, or that the desire of *chanda* feels cruel. But that doesn't really get us closer to anyone's internal feelings. What does "cruel" actually feel like?

"Can we ever know how someone felt in the past?" is the perennial question hanging over the head of every historian of emotion. Usually, the answer is no. But sometimes, just sometimes, the people of the past were thoughtful enough to have been explicit about what they felt. The ancient Hebrews were one such group of thoughtful people. But they also wrote about what feelings are and what they do, so we'll start there.

The Hebrew language, like most languages, doesn't have a good translation for the English word *emotion*. This is mostly because, as you might recall from the introduction, the idea of emotion is a modern English invention. Of course, emotions of various kinds pop up in the Hebrew Old Testament, or Torah, quite a bit. There are many descriptions of the wrath of God, of anger, fire, brimstone, and terrible sorrow. There's also a great deal of love and compassion. But ancient Hebrew emotions were complicated. Feelings now translated as "love," "compassion," and "anger" were not understood by ancient people in quite the same way as we understand them today. Emotions back then weren't seen as psychological phenomena; they were based on the behavior of God—Yahweh—and the rituals surrounding him.

To understand what that means, we must first take a look at God's relationship to emotion in the Torah.

When the writers of Jewish scripture describe Yahweh, they don't portray him as a man with a long gray beard or anything like you'd find on the ceiling of the Sistine Chapel. The most complete description is found in Exodus 34:6–7, in what is known as the Thirteen Attributes of Mercy. These attributes are more of a personality study than a physical description. According to these passages, God is deeply emotional. On the one hand, he is filled with a tender, fatherly, and affectionate compassion (*ra-ḥûm*) combined with a sort of warm-hearted graciousness (*wə·ḥan·nūn*). He's filled with goodness and truth (*we·'ĕ·me ḥe·seḏ ṯ*). Part of that goodness is a deep reliability or loyalty — a promise that he'll have your and your family's backs for thousands of generations. Yahweh also promised that he would suffer our sins for us (*nōśê awon*).[7] Yahweh is "slow to anger" (*'ap·pa·yim 'e·rek*).[8] On the other hand, that doesn't mean he can't get angry. If you fail to repent for your sins, watch out. It's not only bad for you, it's also bad for your kids, your kids' kids, and their kids. Unrepentant sin forced Yahweh to remind his people that

> I, the Lord your God, am a jealous God, bringing the consequences of the fathers' iniquity on the children to the third and fourth generations of those who hate me, but showing faithful love to a thousand generations of those who love me and keep my commands.[9]

In short, God can be very loving and compassionate as long as you do as you're told and apologize properly when you don't. If you sin and don't repent, there will be trouble.

What constitutes a sin is anything that runs the risk of revolting God. The ancient Hebrews believed that Yahweh focused on sin not just because it was intrinsically evil to him but also because it made him feel various types of revulsion. In the Latin Vulgate Bible, this feeling is translated using one word: *abomination* (*abominatio*). We'll come back to that word when we talk about witches later in the book. Modern Bibles sometimes use the word *disgust* as well as *abomination*. Unfortunately, neither one of these English words is quite right. To understand why, it might be worth exploring what modern science thinks disgust is.

A Revolting Science

Full disclosure: my main field of study is disgust. I did my PhD on it; I write about it; I think about it. In much of the world of the history of emotion, if someone thinks of disgust, he or she thinks of me. Or so I'm told.

Some people argue that disgust is universal—a moral-gatekeeper emotion we all share. Fellow disgustologist Valerie Curtis certainly sees it that way. She thinks that we evolved to feel revulsion so we can avoid those objects, animals, and people that make us ill. She calls it parasite avoidance theory, or PAT. According to Curtis, the reason we also find immoral acts repulsive is that they are a sort of contamination. We think people act up because they are infected in some way and that their infection might spread.[10]

But I'm not sure Curtis is right, not entirely. And I'm not alone. The late pioneering neuroscientist Jaak Panksepp didn't think disgust was a universal emotion. He asked, "If we consider sensory disgust to be a basic emotional system, then why not include hunger, thirst, fatigue?"[11] It's a good question and a hard one to answer. What we do know is that various people find various things revolting—and this is a critical feature of revulsion.

For example, just about every culture eats some kind of food that others think of as gross. Scotland has haggis: sheep's stomach stuffed with barley and other ingredients that you don't want to know about.

Casu marzu cheese *(Photograph by Shardan)*

In Sweden, people enjoy surströmming, a herring dish that is fermented for at least six months in a can. The opening of the can releases one of the most potent smells you'll ever experience—trust me. In Sardinia, there's a famous cheese called casu marzu that contains live maggots for an extra bit of protein. Many cultures and quite a few New Age alternative-medicine gurus prescribe urine and feces, both human and nonhuman, as medicine. When it comes to being revolted, we humans are an odd bunch. But it's not just what we find revolting that separates us: the very words and concepts we use do, too.

You don't have to go back in history to see that disgust isn't as universal as you might think. The English word *disgust* refers to the type of revulsion you'll find in most psychology papers, and that's mostly because of the dominance of English-language psychology journals. This form of revulsion is mostly about not looking, touching, smelling, or tasting something because it makes you feel nauseated.

Encountering such a thing would also cause you to make a particular facial expression known in psychology as the gape face. You can imagine what it looks like: wrinkled nose, furrowed eyebrows, lips in a downward curve.

The German word for revulsion, *ekel*, doesn't necessarily invoke nausea. Generally meaning "to move away from or avoid something unpleasant," *ekel* once upon a time could be caused by being tickled. The French word *dégoût* is slightly different again. It's that feeling you get when you've had too much of a good thing. One slice of cake too many; wearing too much perfume; going over the top with the decorations. The differences continue from language to language and culture to culture. It seems to me that we do have some sort of protective feeling of revulsion that stops us from eating, smelling, or even touching a rotting apple. But, as I'm sure you realize by now, there's much more to emotion than a stimulus followed by a response.

God's Revulsion

The ancient Hebrews had several words that applied to the many types of disgust Yahweh experienced. *Shaqats*, or *sheqets*, was closest to the *yuck* of modern disgust. It was what Yahweh felt if you ate or even touched certain unclean animals, including shellfish and pigs.[12] Another type of disgust was the hatred-like and revulsion-like feelings of *toebah* and *taab*. These were caused by people and objects that were ritually unclean or immoral, and they were more about an active dislike than about physical revulsion.[13] Worst of all were the waves of deep anger and disgust called *shiqquts* and the less common *gaal*. The closest we can get to *shiqquts* today is perhaps the sort of feelings we direct toward those who have committed morally repugnant crimes such as pedophilia—a mix of disgust and rage. Yahweh is said to feel *gaal* when he comes across blatant idolatry. Idolatry was just about as great a sin as there was. It wasn't just worshipping a graven image or another god but also doing so in the Temple, in God's sacred space. Those who were guilty of idolatry were beyond repentance; they had to be executed, cleansed from the earth.

To do anything that caused Yahweh to feel these aversions was to commit a sin and risk four generations of punishment. The Ten Commandments are part of a long list of things that triggered God's revulsion. This list stretches across the first five books of the Bible—the Pentateuch. The offenses range from using uneven weights and measures to eating shellfish to wearing two different types of material at the same time to coveting what another person possesses to theft and murder. For the most part, the ancient Hebrew texts only describe what Yahweh felt about things. But as I mentioned above, on occasion they also describe how the Hebrews themselves felt.

To the Hebrews, emotions were visceral. The heart was the seat of the will, the intellect, and certain feelings, such as kindness. The kidneys, *kilyot*, were vessels responsible for the deepest emotions of the soul. It's there where you experience the gut feeling you have when something just isn't right. The liver, *ka'veid*, was the seat of glory and honor. The nose, *aph*, took its name from *anaph*—meaning "breathing heavily" or "snorting in rage." (The phrase I translated above as meaning "slow to anger"—'*ap·pa·yim 'e·rek*—literally means "long nostrils." This is not an insult but an observation that it takes a lot for God's nostrils to get hot.) The womb (*rechem*) was linked to deep compassion, like that of a mother, and not unlike Yahweh's *ra-hûm*. As an ancient Jew you absolutely "felt" emotions, deep within your sinews. But those internal feelings were part of a broader understanding of the body, one that was closely linked to the rules of society. Physical, emotional, and social suffering were all thought of as the same thing.

Ancient Hebrew feelings accompanied ritual practices. Love and warmheartedness were generated by sticking to the law as given by Yahweh—being a useful member of society made you happy. Sin was something you actively desired not to do, because to sin was to flout

God's laws. The feelings those transgressions caused would likely be similar to the types of revulsion people attributed to Yahweh. Whether humans were made in God's image or God was made in theirs, the descriptions given in the Pentateuch would have been a reflection of the emotions ancient Hebrews felt themselves. Those feelings could be controlled through patterns of ritual cleansing. An excellent way to soothe that revulsion—in Yahweh, sinners, and witnesses to sin—was to create pleasing aromas by burning meat and grain in the Temple. Most of the time, asking for forgiveness in the form of an animal blood sacrifice was enough.[14] The bigger the blood sacrifice, the more powerful and long-lasting the forgiveness and the larger the number of people who could accept the sacrifice as being done on their behalf.

Anger in the Temple

To Paul and the first Christians, the Temple sacrifice had been replaced by the blood sacrifice of the Messiah. Jesus, the Messiah—some said the son of God himself—had offered his blood up to Yahweh to reduce his aversion to our sins and so grant forgiveness to all. Whereas the blood of an animal might placate God's wrath for a short amount of time, the execution of his son pacified him for eternity—that is, if you believe that the offering was made for your sake. To the first Christians, Jesus's sacrifice was the equivalent of a ritual atom bomb, an atonement so powerful it could forgive everyone. To Paul, that meant *everyone*, Jew and Gentile alike.

It was that indiscriminate preaching, as well as what he was preaching, that got Paul into trouble during his visit to the Temple. To the Jews, the idea that a man's sacrifice was enough to provide forgiveness for anyone who wanted it was blasphemy. It meant that you couldn't sin; that all was forgiven; that the law meant naught. It's why the people in the Temple at Jerusalem, consumed by the revulsion of *shiqquts* thanks to the idolater in their midst, wanted Paul dead.

Despite his claims to be pure, the Hebrews thought, Paul's liver was empty—he had no honor. In the eyes of the others in the Temple, Paul was a menace. He was a taint, a contaminant, an undesirable. He was actively spreading beliefs that most Jewish people could not accept, and yet there he was, in the Temple of Jerusalem, trying to pass himself off as one of them. It was idolatry. That would not do. Yahweh needed more than the scent of a little barbecue before he could let the defilement of his Temple go unpunished. Paul's presence in the holiest place in the Jewish world no doubt made the kidneys of the others in the room ache with malevolence and their nostrils snort with rage. He was causing full-blown *gaal* in the people around him and therefore, they assumed, in Yahweh. The proper reaction was to remove the dangerous idolater from the Temple and, if possible, from the face of the earth.

Despite the ruckus in Jerusalem, Paul was almost as good at converting Jews as he was at angering them. He understood their emotions, and he was able to connect with them using language and imagery they understood. For example, he once gave a speech in the synagogue in Pisidian Antioch—a town that lay in what is now western Turkey—in which he changed Jewish hearts and minds, or, rather, livers and kidneys. He referenced the long history the Jews had with Yahweh: "For about forty years he put up with them in the wilderness."[15] He engaged his audience's honorable livers when he compared Jesus to other great men from Jewish history: "He raised up David as their king and testified about him." Paul presented Jesus as free from death, incapable of producing the disgust-like sensation of *shiqquts*, because he was "never to return to decay." And he presented Jesus as a figure of mercy (*hesed*) and forgiveness (*nōśê*): "Let it be known to you, brothers and sisters, that through this man forgiveness of sins is being proclaimed to you." After his speech, the book of Acts claims that "many of the Jews and devout converts to Judaism followed Paul and [his friend] Barnabas... urging them to continue in the grace of God."

That his preaching to his fellow Jews worked for the most part was all well and good, but Paul didn't want to convert only Jews. Like other Jewish people, he still believed that all Gentiles were idolaters by definition. He also thought that the blood sacrifice made by Yeshua ben Yosef was so great that even they could be prepared for the impending judgment—but only if they could be convinced that Yeshua's blood sacrifice was also made on their behalf. In Paul's view, the best response to idolatry wasn't capital punishment: it was the offer of a new life. But traditional Jewish arguments were not enough to convince most Gentiles; they didn't understand the emotional landscape on which these views were built. Paul had to take another tack.

Paul the Stoic

This story begins seven years before Paul's near-death experience, with a guffaw, then a splutter, then a roar. It continues with the sound of laughter echoing down from the Hill of Ares, near the Acropolis, in Athens. The hooting and cackling likely reached the Parthenon, bouncing off the beautiful white and pink marble before careering down the hill to the marketplace below. This wasn't the friendly sort of laughter you get when someone's told a joke. These were hysterics of the nastiest kind: the mirth of mockery; the hilarity of scorn.

On the top of the Hill of Ares sat the ancient home of Athenian democracy—the Council of the Areopagus—and standing in that court was Paul. Before the laughter began, he'd been well respected enough for the members of the ancient Athenian court to take the time to listen to what he had to say. But as was so often the case outside the Holy Land, they thought what he'd said was ridiculous. Paul had just told a room filled with Greek intellectuals that a man had risen from the dead. Not in a spiritual or metaphorical sense. He'd claimed that this man had literally gotten up and wandered around as if he'd never died. The Greeks thought that was hilarious, and that's when their laughter echoed through the court, the Parthenon, and the marketplace below. Paul thought he understood his audience. He was, mostly, wrong.

Although Paul was raised by parents who tried to shield him from Greek culture, he appears to have known a great deal about it. In fact, he was something of an expert in it. This is not surprising. As a young Jew in a Greek-speaking province, every debate he'd ever had with a non-Jew would have been with a follower of Greek thought. As an adult, he could recite Plato and Aristotle, and he was extremely well versed in the most popular Greek philosophy of the time, Stoicism. This is why Paul was confident he could convert even the most intellectually elite people in the Roman Empire. It's also how he ended up standing before the court of the Areopagus on the Hill of Ares, in Athens.

Paul had been invited to the court because he'd been preaching in the marketplace, or *agora*. Well, not so much preaching—that top-down method wasn't working. Instead, he decided to model himself after Socrates, standing at the edge of the *agora* and asking questions of passersby, slowly digging into their beliefs until they became convinced of Jesus's resurrection. Paul's debating had become popular with two philosophical groups that he knew well—the Stoics and Epicureans. The Stoics, whose belief system was by far the dominant one in the Roman Empire at the time, wanted to know more. They politely invited Paul to make his case in the court. Once inside, he availed himself of his great intellect and his deep knowledge of Greek philosophy. He knew he wouldn't convince a room filled with scoffing academics that his message was the truth, but he wasn't really aiming it at them. The Bible is filled with accounts of Paul and other apostles preaching to large numbers, knowing they might only reach a few on the periphery. Paul was likely targeting the other onlookers—the seekers, the searchers, the ones for whom Stoicism wasn't enough. To reach them, he knew he had to grasp their hearts as well as their minds. He knew he had to risk their laughter.

Before we go any further and find out what Paul said to cause such hilarity, let's take a short detour into what Stoicism was and where it likely came from.

When most people think of Stoics, they imagine someone a bit like Mr. Spock from *Star Trek*: calm, completely emotionless, basing all his decisions on cold logic. But the Stoics weren't like that at all. Well, not entirely. Unlike Spock and the Vulcans, the Stoics were allowed to experience emotions, even use them when making decisions. But they had to be the right kind of emotions. Stoicism was more than just a philosophy; it was a way of life. To be a Stoic took dedication and focus—it was every bit as central to a person's life as religion. At the base of it was a take on Plato's emotional regime and the idea that in order for a person to be virtuous, he or she had to keep feelings under control. But the Stoics took it a step further and tried to work out how to control their feelings in order to live a better, happier life.

To live a happier life, you had to understand that all living creatures are drawn toward that which benefits them and are repelled by that which causes harm. But this is a bit like the *eros* form of good: sometimes a bad thing can lead to a greater good—such as the amputation of a diseased limb that might otherwise kill you. But to a Stoic, the only things that are really beneficial are those that are virtuous, regardless of how pleasurable they are. Everything else is a "matter of indifference" (*adiaphora*). Obsessing over how much money you have, how healthy you are, or which god you should worship won't lead to virtue. That doesn't mean you should ignore anything that's not important; it just means that you shouldn't make a production out of it. You should live the life you find yourself with, do what is expected, and not fret about the irrelevant details. For example, if you find that fate has made you a powerful ruler, there are elements of your role that are matters of indifference. These might include protecting your borders, expanding your power, ordering executions, and so on.

If a focus on doing your duty regardless of how you feel sounds a bit like *dharma*, then what you're about to read will cause even more déjà vu. Stoics believed that to become genuinely virtuous you had to control your desires and only give assent (*sunkatathesis*) to those that are genuinely beneficial to you or society. To achieve this assent required adherents to follow a complex system of mathematical logic that I refuse to bore you with here.[16] Suffice it to say that you can't just do things on a whim, because that will lead to *pathē*—those disturbances we last visited in chapter 1, which will in turn lead to sorrow and suffering. If you give assent to only what is truly beneficial, then you will experience the right kind of emotions—*eupatheia*.

The Stoics identified four core emotions:

1. Good things that are going to happen in the future can produce either the *pathē* of desire or the *eupatheia* of wishing. Desire is terrible, because it can never be fulfilled. A bit like the Buddhist notion of clinging, to desire produces a hedonic longing that focuses on material gains. Wishes are also a type of hedonic desire, but they don't always come true, and everyone is aware of that. Unfulfilled wishes are less potent than unfulfilled desires.

2. Good things happening now can create either the *pathē* of pleasure or the *eupatheia* of joy or bliss. Pleasure that comes from fulfilling hedonic desires, again, leads to unfulfillment and so to sorrow. Joy or bliss, on the other hand, are by-products of wishing to be a good Stoic—the results of a motivational desire.

3. Bad things that might happen in the future triggered the *pathē* of fear or the *eupatheia* of caution. Caution is the product of thinking logically about danger and working out what to do. Fear stems from a hedonic desire to get away from something, or fight it, without thinking first. It's far from rational and can lead to more harm than good.

4. Bad things happening now can cause the *pathē* of sorrow and suffering. There isn't a type of *eupatheia* for bad things

happening now, because if you are controlling your thoughts and feelings appropriately, you won't ever feel sorrow.

A Stoic's life's work was to focus on giving thought priority over feeling until it became natural to stop and think about a situation before taking action. Some Stoics—such as Roman emperor Marcus Aurelius—became so good at thinking before feeling that even sex turned into something to ponder rather than lust after. The emperor described sex as "no more than the friction of a membrane and a spurt of mucus ejected."[17] Once you'd reached Aurelius's levels of mastery, you could get beyond *pathē* and free yourself from sorrow, fear, and simple pleasures. The ultimate goal of Stoicism was a state of inner calmness or bliss, *ataraxia*.

I know what you're thinking: *ataraxia* sounds a bit like *nirvana*. Well, you are far from the first to have noticed that. Plenty of historians have suggested that it's a bit more than coincidence. The parallels are there for all to see, and although it's entirely possible that the Greeks influenced Buddhism or that both got their ideas from Persia or China, it is most likely that the Greeks borrowed this particular idea from the Buddhists.[18] There was a moment in ancient history when the Greek and Indian cultures touched. If you remember, when we last encountered Alexander he was standing in India, contemplating going home.[19]

One of the people who had accompanied Alexander to India was Pyrrho of Elis. He supposedly met some "naked wise men" (*gymnosophistai*) there.[20] Whether these wise men were Hindu or Buddhist, or perhaps something else entirely, we can't be sure. Regardless, after speaking with them, Pyrrho returned to Greece with the notion that a state of inner calmness was the ideal life goal. He built an entire system of philosophy around it. He thought that if you believe only that which you experience and reject everything else, you can stop fretting about the small stuff and relax. That was Pyrrhonism, or Pyrrhonian skepticism. Stoicism, though somewhat different, was cut from the same cloth: a search for *nirvana* — sorry; I mean *ataraxia*.

While Stoicism, Pyrrhonism, and other calmness-focused philosophical lifestyles that sprang up around the third century BCE might well be Greek-flavored offshoots of Buddhism, I have to stress that no one really knows for sure. The primary source we have for Pyrrho's encounter with naked Indian wise men comes from Diogenes Laërtius's *The Lives and Opinions of Eminent Philosophers*. This work was written some six hundred years after the event. All we really know is that Pyrrho suddenly went sort of Buddhist in his thinking, as did other schools of Greek philosophy, including the Stoics. Regardless of who influenced whom, Saint Paul understood the Stoic take on emotions, and that brings us back to his address at the Areopagus, which caused so much laughter. What did Paul say that was so funny?

The Pauline Joke

It wasn't exactly a stand-up routine. Paul began by paying what to him was a compliment to his audience: "People of Athens! I see that you are extremely religious in every respect."[21] He then told them about an altar to an unknown god he spotted on his way to the court, using it to suggest, in language they understood, that they really didn't know God at all: "The God who made the world and everything in it — he is Lord of heaven and earth — does not live in shrines made by hands."[22] The idea that God wasn't concerned with irrelevancies such as statues and altars was his first appeal to the Stoics in the room.

So far so good. He continued, "Neither is he served by human hands, as though he needed anything, since he himself gives everyone life and breath and all things."[23] Again, this would appeal to a Stoic. If there is an all-knowing, all-powerful being, why would he need humans to do things for him? That would be wasting energy on what, to God, would be matters of indifference.

Paul marched on: "From one man he has made every nationality to live over the whole earth and has determined their appointed times and the boundaries of where they live."[24] Here, Paul is still not straying too far from Greco-Roman beliefs. They, like ancient Hebrews, believed that all humanity was made by one god—in the Greeks' case, Prometheus—from clay and then spread across the planet.

Paul delved deeper and more obviously into Greek thought: "He did this so that they might seek God, and perhaps they might reach out and find him, though he is not far from each one of us."[25] This line was a setup, enabling Paul to drop some classic Greek philosophy into the conversation: *"For in him we live and move and have our being,* even as some of your own poets have said, *'For we are also his offspring'* " (my emphasis).[26] The first part of that sentence was a quotation from a poem by a philosopher even older than Socrates, Epimenides. The second was from the Stoic philosopher Aratus. Paul was coming into the Greeks' world, talking to them on their terms.

Then came the zinger: "Since, then, we are God's offspring, we shouldn't think that the divine nature is like gold or silver or stone, an image fashioned by human art and imagination."[27] The Stoics would certainly be on board with this. Stone was just stone; gold just gold. Impermanent, irrelevant. Not something you should desire if you want inner peace. You can imagine the Stoics in the room leaning forward, eager to hear what this man had to say.

Then Paul made a mistake. In a room full of Greek Stoics, he veered too far into his own belief system. "Therefore," he said, "having overlooked the times of ignorance, God now commands all people everywhere to repent, because he has set a day when he is going to judge the world in righteousness by the man he has appointed."[28] This came from the Jewish mind-set. It was about asking forgiveness for causing God revulsion. Paul might have thought this a good segue, saying to the Stoics that the only thing that isn't a matter of indifference is the will of God. He seems to have been suggesting that to be truly virtuous and avoid *pathē* and sorrow, you need faith.

Paul thought he had proof: "He has provided proof of this to everyone by raising him from the dead."[29]

With that, he blew it. The claim of the dead man walking cost him all credibility in the eyes of his audience. Then, as now, people simply didn't shake off being executed and go about their business. Surrounded by uproarious laughter, Paul had no choice but to leave. But all was not lost. Not everyone laughed. Some wanted to know more.[30] His words had clearly connected with at least a few hearts and minds. One man was particularly interested. A Neoplatonist—that is, someone who adhered to an updated version of Plato's beliefs—named Dionysius wouldn't stop asking Paul questions all the way down the hill. He became a committed Christian. Another convert was a woman named Damaris. There may have been more; we don't know. We do know that Paul had used Greek thought—particularly the Greek view of emotion—to make his case. He had claimed that the way to *eupatheia* was through faith in Christ. If you weren't laughing at the dead man walking, you might be fascinated by this new way to bliss.

The Feelings of Paul the Christian

Saint Paul's understanding of emotion had an incalculable effect on the world. Much of modern Christianity comes from the fusing of two ideas about feelings started by Paul. On the one hand, there's the ancient Hebrew idea that you should avoid sinning so as not to cause God aversion. If you do sin, so long as you have faith, your transgression can be forgiven by accepting a blood sacrifice to Yahweh. This sacrifice is now mostly understood as the suffering and crucifixion of Christ for our sins. The blood-sacrifice element is all but forgotten. On the other hand, Paul tapped into the Stoic idea that happiness can only be gained by focusing on what is genuinely virtuous—in this case, wishing, if not desiring, that your sins will be forgiven.

That, in essence, is why there are now about 2.4 billion Christians around the globe, making it the world's largest religion—because Saint Paul managed to meld how he felt about Jesus as a Hebrew with how he felt about Jesus as a Stoic.[31] It allowed more Greek-speaking people to believe. It allowed the word to spread and, eventually, it allowed emperors to take up the cross.

But there is another man who had almost as much impact on the Christian worldview as Paul did. He achieved this with his ground-breaking ideas about one emotion in particular: love. To explain, I'm going to use an event from history that doesn't seem to have much to do with love at all—the cruel and violent episodes known as the Crusades.

Four

Crusader Love

In 1095, in the city of Clermont, France, the pope was getting ready to deliver an address before the Council of Clermont—a meeting, called at his request, that gathered together lords, priests, and gentlemen. The speech he'd planned for this event was to be the most important of any given during the Middle Ages. I say *the* pope, but Urban II was just *a* pope. His old mentor, Pope Gregory VII, had annoyed a great many influential people by suggesting that only the Church, not royalty, ought to have the power to appoint priests and bishops. The Holy Roman Emperor, Henry IV, disagreed and was excommunicated. That kicked off nearly fifty years of violence in what has become known as the Investiture Controversy. Quite early on, Henry IV decided to do some excommunicating of his own. He pronounced Gregory deposed and installed his own man—Guibert of Ravenna—as Pope Clement III.

As a protégé of Gregory VII, Urban had always been on his side. But the schism bothered him. It wasn't the only one that did. In 1054, a series of differences of opinion had split the Christian Church in two. Some of the arguments were seemingly mundane, such as those over which kind of bread should be served at the Eucharist. Others were significant theological disputes that stretched back centuries, such as whether the Holy Spirit proceeds from the Father and the Son or just from the Father. In other words, was Jesus actually God or just the son of God? In the end, the Catholic and Orthodox churches went their separate ways. Urban thought he might be the man to unify them once more.

Fortunately for Urban, he'd just received a letter that might allow him to kill three birds with one stone. First, he could pull the bishops who supported his rival, Clement, back to his side. Second, he could give the nobles of Europe something to do other than interfere with the reforms he'd inherited from Gregory. Third, he might be able to heal the wounds separating the Catholic and Orthodox faiths.

Emperor Alexios I Komnenos, ruler of the part of the world in which the Orthodox Church dominated—Byzantium—had written to Urban asking for help. Ever since the Seljuk Turks had taken control of Jerusalem, in 1071, Christian pilgrims had been denied access to holy sites. Worse still, those who tried to gain entry were met with intimidation and violence. On top of that, the Seljuks were slowly encroaching on Byzantine territory.

Urban saw the speech he'd been scheduled to make at the council as an opportunity to unite Christendom against a familiar foe. He had to tap into powerful expressions of emotional persuasion—send a message to rouse the rabble, to call people to arms, and march on the Holy Land. And it worked. Pope Urban II managed to kick off the First Crusade.

There are a lot of moments in history that we historians find puzzling. How did the Etruscan language work? Who went around burning Mediterranean cities to the ground in 1200 BCE? But one of the strangest occurrences in history was the Crusades. In the halls of history, the precise meaning of the term *Crusade* is disputed, but I'm taking it to mean what most people take it to mean: burly knights on horseback, clad in armor, riding off to the Holy Land to overthrow— or, rather, oppress and slaughter—the forces of Islam.

One way to think of the Crusades, and a way that many crusaders themselves thought of them, is as a series of violent pilgrimages: journeys to the lands at the heart of their faith for the glory of God. There were seven distinct Crusades that took place over the course of 176 years, beginning in 1096.* There were some incredible successes— from the crusaders' point of view, at least. But there were also some abject failures, including the Fifth Crusade (1217–21) and Seventh Crusade (1248–54). Sometimes, Crusades ended in a truce, as they did when King Richard I made a deal with Saladin, the first sultan of both Egypt and Syria, in the Third Crusade (1189–92) and when a series of political disputes was settled through negotiation in the Sixth Crusade (1228–29). The Fourth, or Unholy, Crusade (1202–4) included entire crusader armies being excommunicated after they went rogue, stole boats from the Venetians, and invaded Constantinople rather than the Holy Land. It's worth pointing out that the excommunication was lifted when that invasion of Constantinople proved successful. The Crusades were varied and odd, and the reasons so many of them occurred remain perplexing to this day. But one aspect I think is a little less perplexing is why they were instigated in the first place. It was really all about love. That's right, *love*.

What Is Love?

Modern neuroscience likes to break love down into its constituent chemicals and stages. First, there's *lust*, kicked off by the hormones that live in our sex organs—testosterone and estrogen. These chemicals are produced when the part of our brain called the hypothalamus

---

* Or eight? Or nine? Unless eight and nine were seven? But maybe there weren't any— they were just part of ongoing European wars of expansion. Or maybe there were many more, because Europe had been attacking Islamic nations long before 1096. Or maybe the first two didn't count because they didn't call them Crusades at the time. It generally depends on which historian you ask and what mood he or she is in.

decides it's time to breed. It has a pretty obvious evolutionary function: reproduction.

Then, if you like the person you lust after, the hypothalamus kicks out another chemical—dopamine—to trigger *attraction*. Dopamine is known as the reward chemical. It's the one that's released when you do something that makes you feel good, such as take drugs, skydive, or eat chocolate. When high levels of dopamine are blended with the fight-or-flight hormone, norepinephrine, you might become excited and filled with energy. According to anthropologist Helen Fisher, the reward parts of the brain light up like a Christmas tree in an fMRI when people see pictures of individuals they're attracted to.[1]

Finally, there's *attachment*, which mostly relies on oxytocin. Oxytocin is released during bonding—specifically childbirth, breastfeeding, and sex. It's also been measured in pets—even seemingly love-deficient cats like my Zazzy—when their owners return home.[2] But oxytocin is essential. It moves you from a state known as *limerence*—that lustful, obsessive period at the beginning of a relationship—to the bonding associated with long-term relationships. But it's not just a love chemical.

The best way to think of oxytocin is as the fuel in what I like to call a belongingness engine. Belongingness is important to hypersocial species such as humans. Attraction to one another—feeling like we belong within a group—turns out to be almost as essential to human survival as drinking water. Major psychological harm, even death, can result when people are kept isolated from other like-minded people.[3] The belongingness engine so important to our survival needs much more than a release of oxytocin in the brain. We have to interact with the people we care about, experience physical contact, talk, laugh, take part in shared activities. The more of that we do, the stronger our sense of belongingness—or our love—for one another. But that's not always a good thing. Having strong connections to one group while being only weakly attracted to others can lead to violence. Feeling too much attraction can also be dangerous, especially if oxytocin is sloshing about in the system while the object of that attraction is unavailable or the love is unrequited. It's a state of affairs that can cause anything from loneliness to dangerous obsessions.

Modern philosophy treads a different path from science when it talks about love, albeit one that's consistent with science. Some philosophers describe love as a union between people—people who are attracted to each other or bonded in some way.[4] Others describe it as "robust concern"—that is, actively caring about someone else's well-being, perhaps even more than your own.[5] This also requires some type of attraction and bonding. Love is also thought of as a process of attaching some sort of value to a person. This is known as an "appraisal of value."[6] The more you value someone or something, the more you love her or him or it—or, perhaps, the more love neurochemicals are released.

But there's another side to this discussion. Some philosophers argue that love isn't an emotion at all. Instead, it's an *emotion complex*— the sum of a range of feelings that you attach to something or someone over time. That might be a mix of lust, attraction, attachment, care, perhaps even frustration at times. In this view, that complicated bag of feelings you have for someone you care about is what builds your love for that person. This view explains why no two loves are the same. I don't love my wife the same way I love my mom or Zazzy because the complex of emotions is different in each case.

But none of these views is mutually exclusive. We love people because they have value to us, because we care for their well-being, and because we want to be with them. And that love could be the result of a mixed bag of feelings that exist because specific neurochemicals are doing their jobs.

But the type of love the crusaders were drawing on wasn't found in modern science or philosophy. Well, not exactly. It was a type of love that had been described just over seven hundred years before Urban made his speech, one that formed the basis of an emotional regime that dominated Europe at the time of the Crusades. To understand that emotional regime, we need to go back to the man who constructed it: Saint Augustine.

Augustine Wept

In August of 386 CE, a man sat alone in his chambers in Milan, weeping. A flood of tears rolled uncontrollably over his desk because he had just read something that had changed his life and, arguably, ours, too. His name was Aurelius Augustinus—after his canonization, in 1298, he would be known as Saint Augustine.

Ever since leaving his native North Africa, and indeed for some time before, Augustine had adhered to a religion called Manichaeanism. Though it's now defunct, Manichaeanism once rivaled Christianity and Islam in its popularity. Its influence stretched from the Atlantic shorelines of Europe to the Pacific coast of China. The Manichaeans believed that there were two godlike forces at work in the world—one bringing light and one bringing darkness. These were always doing battle, and both the earth and human souls were part of the struggle. That's why we have night and day, sunshine and rain; why we can feel good when we've done something wrong and terrible when we've been good. It's all about balance. As a young man, Augustine had taken great joy in stealing fruit for the sake of it, so Manichaeanism made sense to him back then.

The problem for Augustine in 386 CE was that it didn't make sense anymore. He couldn't work out why we sometimes feel good when we do bad things. Joy shouldn't have accompanied his theft, and he knew it. He'd never really been a staunch adherent of the Manichaean faith, remaining on the lowest rung of its strict hierarchy as an "auditor" despite being among its best-read disciples. He studied harder than the people around him, absorbing the minutiae of Greek and Roman philosophy as well as the texts of his own faith. And that was part of the problem. Through his research, he'd become enamored enough with Plato to regard himself as a Neoplatonist, and he'd come to wonder whether there wasn't another path to God.[7]

By 386 CE, when we find him weeping alone in his chambers, he was deep in the throes of a crisis of faith. A fifteen-year-long relationship with the mother of his child ended abruptly when his own mother commanded him to marry a rich young heiress. To make matters worse, the new fundamentalist Christian emperor—Theodosius I—had ordered all Manichaean monks to be killed on sight. Augustine had to deny his belief whenever he was questioned about it. His life was all darkness, no light. There was no balance to be found.

On that fateful day in 386 CE, Augustine's good friend Pontitianus had come to visit. Pontitianus noticed that Augustine had a manuscript of Saint Paul's letter to the Romans rolled up on his desk. This made him smile, because he, like his emperor, was a devout Christian and keen to find new converts. Augustine left his chambers with his companion to meet up with some of their friends and share dinner and conversation. At one point, Pontitianus began to read a book out loud. It was about the conversion of another man—Antony of the Desert. Augustine liked what he heard, and something about it resonated, getting him to think about his own life. To use his own words:

> Thou, O Lord, while he was speaking, didst turn me round towards myself, taking me from behind my back where I had placed me, unwilling to observe myself; and setting me before my face, that I might see how foul I was, how crooked and defiled, bespotted and ulcerous.[8]

In other words, Augustine became *revolted* with himself, with his behavior, with his lifestyle, with his faith. Most of the religions that sprang up around that time still carried with them the sense of repulsion that was central to Hebrew belief. Augustine remonstrated with himself for once praying to the God of light, "Give me chastity and continence, only not yet," a prayer he'd seen as giving him a green light to spend his life in pursuit of carnal pleasures.[9] Augustine had desired sex; he had desired food; he had desired physical beauty. He was "gnawed within, and exceedingly confounded with a horrible shame."[10] Augustine began to wonder why he so easily gave in to his sinful nature. He asked:

> Whence is this monstrousness? and to what end? The mind commands the body, and it obeys instantly; the mind commands itself, and is resisted. The mind commands the hand to be moved; and such readiness is there, that command is scarce distinct from obedience. Yet the mind is mind, the hand is body. The mind commands the mind, its own self, to will, and yet it doth not.[11]

Before Augustine knew it, he was on the verge of tears. He made his excuses to his friends, already choking up as he bade them farewell, and returned to his chambers. There he stayed for some time, howling in grief and shame. All the pressures of his life closed in on him at once — the sin, the lust, the selfishness. He wanted to be different. He desired change.

It was then that he heard a voice. It wasn't a prophetic, powerful voice. Nor was it what Paul heard — the soothing yet somehow violent tones of Jesus himself. What Augustine heard was the soft, calm voice of a child chanting. The boy said, "Take up and read; take up and read." Augustine realized that he still had the scrolls containing the books of the New Testament on his desk, and he decided to do as the child suggested. He unrolled one and read the first thing he saw: "Go, sell your belongings and give to the poor, and you will have treasure in heaven. Then come, follow me" [Matthew 19:21]. Augustine exploded into a flurry of excitement. The instruction touched him. It made profound sense. He decided to take up another of the scrolls on his desk, this time the letter to the Romans, and rush back to his friends. On reaching them, he chose another passage at random:

> Let us walk with decency, as in the daytime: not in carousing and drunkenness; not in sexual impurity and promiscuity; not in quarreling and jealousy.[12]

At that moment of pure emotional release, Augustine became not only a Christian but also one of the most influential philosophers and theologians in history. His works would go on to inspire Christian thinking for the following millennium and more as he developed and expanded crucial doctrines—the original sin of Adam and Eve; the importance of the Virgin Mary; the concept of human free will. Most important, he developed the idea of grace. To Augustine, the purpose of a good Christian was to seek God's grace by rejecting the sins that had so tempted his younger self. Instead, he thought, one should focus on the higher, inner self of the soul. This might seem like nothing new—yet again, we find a man advocating that we learn to control our carnal desires and instead crave a higher purpose. He was another Plato, striving for *eros*; another Buddha, struggling toward *nirvana*. But the difference was that the focus for them had been desire—how to use it, suppress it, challenge it. Saint Augustine, like the Beatles, believed that love is all you need.

All You Need Is Love

As I've said, Augustine was influenced by the Greeks, especially Plato. He believed in Plato's three-part soul. Like Plato, he thought that the rational soul represented our pure form, a noncorporeal, almost supernatural, perfect version of ourselves. But Augustine had his own take on that structure. He simplified Plato's thinking a little, influenced in no small part by his years as a Manichaean. He suggested that there were two parts to the soul, one dark and one light. The bit that couldn't think, the bit that acted instinctually, was our dark outer soul—our flesh. The bit that could think, the bit that acted through judgment and deliberation, was our light inner soul. And it was up to us which of these souls would guide us. Unlike the Manichaeans, Augustine didn't think the split between the light and dark parts of the soul was caused by two battling gods. Instead, we caused the split ourselves when we disobeyed God in the Garden of Eden. When Genesis 1:27 says, "So God created man in his own image; he created him in the image of God; he created them male and female,"[13] the "image" in question is the light inner soul. It's the bit of God in all of us. Our true selves. The part that was tarnished after Adam and Eve disobeyed God's explicit command and tasted the apple of the Tree of the Knowledge of Good and Evil. Their subsequent banishment from paradise didn't refer to a mere physical place. It was a spiritual place, a dark outer self created by the act of original sin, which imprisons us in our flesh.

Like Paul before him, Augustine thought one ought to focus one's mind on faith in Christ and his sacrifice. But Augustine didn't buy into the Stoic idea of evaluating your emotions. He agreed that feelings were chaotic and violent, able to lead you astray. But to him, the trick was to find a way to pay attention to your inner self until you can once again see the vision of God buried deep inside you. This is how you could reject the chaos of the corporeal world—something he called the City of Man—and fixate on the incorporeal world to come: the City of God.

Augustine understood that this wasn't easy, but he had a method: you had to try to behave as Jesus did. When Augustine heard the voice telling him to read, he felt overwhelmed with love. Through that love, he gained knowledge; that was his reward. To Augustine, Jesus was the embodiment of love and wisdom. God the Father was memory and remembering. The Holy Spirit was the will, but Jesus, the Son, was wisdom, love, and understanding. It was through this wisdom that Jesus, like Socrates, was able to control his spirit and his will even while being put to death. Augustine believed that Jesus was a reminder of our true inner selves, made in the perfect image of God. The key to Jesus's perfection was his love for all and the way he expressed that love, through an act of extraordinary grace—his crucifixion.

Augustine thought that the search for God was an emotional one, driven by the right sort of love. It was all about the Golden Rule:

> Love the Lord your God with all your heart, with all your soul, with all your mind, and with all your strength...Love your neighbor as yourself. There is no other command greater than these.[14]

Love, to Augustine, was the passion described by Aristotle as a gateway to truth—in Augustine's case, the grace and truth of God. But just as the soul was divided into two parts, an inner and an outer, he thought there were two types of love.

The first and more common type of love was, in Latin, *cupiditas*, or self-love. This is not all that different from the kind of desire that previous religious thinkers thought we should avoid. It's the selfish, lustful, prideful type of love that mistakes earthly wants for truth and beauty. It's a need that can never be fulfilled. It was this sort of love, he believed, that caused the Fall in the first place; it is the source of all sin.

The second form of love—the love that Jesus was talking about in his Golden Rule—Augustine called *caritas*, sometimes translated as "charity." This is a love of God and of doing things in the service of God. This second form of love itself contained a distinction, one that was central to Augustine's understanding of emotion. First, there was *frui*: things that should be "enjoyed" for their own sake. This applied to God and only God. You were to love God for his own sake because God is love. *Frui* is the route to heaven and to God. According to

Augustine, the aim isn't heaven per se. The aim is to reach God. He is the *finis*, or *telos:* the end; the goal. He just so happens to be in heaven.

The other is *uti:* this is loving something or someone as a way to get to God and heaven—love as a tool. The love you have for your neighbors is this sort of love. Loving your neighbor—or even your enemy— is also done in the service of God because it is, or at least should be, unselfish. This might seem a bit harsh, using your neighbors as some type of tool to reach God and heaven. But it's more a matter of using your love for others as a way to reach God. You don't love them because you want to but because God wants you to, in the way he believed that Christ had used his crucifixion to bring everyone closer to him. To Augustine, when you love your neighbor, you're loving the sacrifice Jesus made for you.[15]

Most important, this *uti* sort of love is the way to find your inner self by focusing on your love of God and on the next world rather than this one. Through love you will learn to control your will, as Jesus did. Love can take you on a path to wisdom and the ability to see the vision of God embedded inside each of us.

The relationship between love and wisdom might seem a little strange. It comes, once again, from Augustine's reading of Plato. Plato understood love as a ladder. The bottom rung was a love of beauty and the physical—lust, you might say. In the middle of the ladder was a love for all parts of a person's soul, what modern science might call affection. Then, at the top of the ladder, was what Augustine would call *frui*. That's love not for the physical but for wisdom, virtue, and the divine—a love of genuine good. It's the love that Plato thought could guide you to *eros*. Some philosophers even argue that *eros is* love, but to me, it refers to the place you're going rather than what gets you there.

Still, for all that complexity, Augustine's love was simpler than Plato's ladder. Plato separated each of his types of love by rungs on that ladder. You could find yourself on a rung in between them, where you love a person's body and soul, but maybe one more than the other. By contrast, Augustine adapted Plato using his understanding of God. He stripped out the nuance and focused only on the extremes, making love binary: *caritas* and *cupiditas*—light and dark.

That binary process didn't stop at love. If love is an emotion complex, it's one that, to Augustine, separated all emotions into light and dark. But this didn't mean that emotions were either good or bad—to Augustine, every emotion had both a positive and negative manifestation. One takes you to the City of Man, and one leads back to the City of God. Anger can be destructive, like the uncontrollable rage that leads to murder, but it can also be loving and righteous, like Jesus's anger at the money changers defiling the Temple in Jerusalem. Fear can be debilitating or lovingly protective, guiding you away from sin. Selfish sorrow for the loss of worldly possessions can be replaced by loving sorrow on behalf of the sinful and remorse for having sinned yourself. To Augustine, and to many before him, emotions weren't intrinsically positive or negative; their moral value was determined by how they were used. Any feeling could be good if it was used in service of God, and likewise, any feeling could be sinful when it was used for personal gain.

Even today, most branches of Christianity put Augustinian love at the center of their faith. The idea that God is love comes straight from Augustine. Augustine is also the source of the idea that Jesus's crucifixion was not a blood sacrifice, as the early Christians believed, but an act of pure grace and love—an act of forgiveness by God, who understands how difficult it is to control the will and who came to earth to demonstrate how powerful the will can be, even in the most extreme circumstances. Although the process of accepting Jesus described by many Christians isn't quite like the meditation and inner journey suggested by Augustine, it's still supposed to be about finding something greater than the self. Augustine's understanding of emotion helped build an emotional regime that came to encompass all Christendom.

The way good Christians were supposed to behave, express their feelings, even pray, had its roots in Augustine's writings. Augustine's love really did change the world. But not always for the better.

Crusading as a Use of Love

Seven hundred years after Augustine's conversion, Pope Urban stood before the packed hall at Clermont, filled with dozens, probably hundreds, of the most powerful and influential people in Europe, including archbishops, abbots, knights, and noblemen from across the region. If he was going to make an impact, now was the time to do it. Augustine's ideas about love and emotion still dominated Christianity at the time, and Urban, a skilled rhetorician, knew how to use them. He began his speech.

> Most beloved brethren: Urged by necessity, I, Urban, by the permission of God chief bishop and prelate over the whole world, have come to these parts as an ambassador with a divine admonition to you, the servants of God.[16]

At least, that's what the version written by Fulcher of Chartres claims he said. There are several records of this speech, each set down by someone supposedly present at the time. In reality, it's likely that the majority are retellings of a small number of original accounts that have been embellished to make a point. That doesn't mean that they are all completely different, though. Almost every account of Urban's speech has him referring to his fellow Christians as "beloved brethren."[17] The words occur not only in the version of the address above but also in accounts by Robert the Monk, Guibert of Nogent, and Balderic, archbishop of Dol.

That the "beloved brethren" term was used as a way to get everyone on the same page in every account is important. Urban and his chroniclers were tapping into the crowd's brotherly, *uti* love for fellow Christians who were up against a common enemy. The most explicit example of this is found in the account by Balderic of Dol. After listing the horrors inflicted by Islamic forces on his fellow Christians living at the edges of the Byzantine Empire—they were flogged, driven from their homes, enslaved, robbed of their churches, and so on—Urban is said to have addressed the crowd directly:

> You should shudder, brethren, you should shudder at raising a violent hand against Christians; it is less wicked to brandish your sword against Saracens. It is the only warfare that is righteous, for it is charity to risk your life for your brothers.[18]

The word used in most of the primary sources for *charity* was *caritas*—that Augustinian right sort of love. But Urban and his chroniclers were also tapping into the direct and powerful *frui* love you should feel for Christ himself. Robert the Monk had Urban use this notion to pry people away from those they loved on earth:

> But if you are hindered by love of children, parents and wives, remember what the Lord says in the Gospel, "He that loveth father or mother more than me, is not worthy of me...Every one that hath forsaken houses, or brethren, or sisters, or father, or mother, or wife, or children, or lands for my name's sake shall receive an hundredfold and shall inherit everlasting life."[19]

The chance of everlasting life in the presence of God was the key. Linking it to a *frui* love for Christ and fellow Christians was powerful.

Much of the crusader rhetoric tapped into *uti* love for the Holy Land itself. The ever-polemical Balderic of Dol echoed Psalm 79:1 by having Urban say:

We weep and wail, brethren, alas, like the Psalmist, in our inmost heart! We are wretched and unhappy, and in us is that prophecy fulfilled: "God, the nations are come into thine inheritance; thy holy temple have they defiled; they have laid Jerusalem in heaps; the dead bodies of thy servants have been given to be food for the birds of the heaven, the flesh of thy saints unto the beasts of the Earth. Their blood have they shed like water round about Jerusalem, and there was none to bury them."[20]

This *uti* love for the Holy Land wasn't just a legend built up by Crusade writers. Islamic accounts of the Crusades put similar words into the mouths of crusaders. Writing about the Islamic reconquering of Jerusalem in 1187, Persian scholar Imad ad-Din al-Isfahani heard terrified crusaders get ready for a final battle with the words:

We love this place, we are bound to it, our honour lies in honouring it, its salvation is ours, its safety is ours, its survival is ours. If we go far from it we shall surely be branded with shame and just censure, for here is the place of the crucifixion and our goal, the altar and the place of sacrifice.[21]

The impetus for crusading seems to have been a deep sense of *uti* love in the Augustinian sense. The problem is, Augustine didn't mean "love only thy neighbors whom you agree with," and that raises a question about how an eleventh-century man might reconcile violence against others with neighborly love. Thankfully, at least from the crusader's point of view, Augustine also had an answer for that in his concept of a just war.

Augustine saw war as an act of correction, a bit like disciplining a child who has misbehaved. He wrote:

They who have waged war in obedience to the divine command, or in conformity with His laws have represented in their persons the public justice or the wisdom of government, and in this capacity have put to death wicked men; such persons have by no means violated the commandment, "Thou shalt not kill."[22]

As long as you are fighting for the right reasons—that is, for God—and not for personal gain or hatred, then the war is just. More than that, it can be an act of *uti* love. Killing a sinner is to remove sin from the face of the earth, and that, to Augustine, was a good thing. It was also a good thing for the crusaders.

Using God's Love

But there's still a problem. The Crusades were clearly not only an act of *uti* love for God, neighbors, and the Holy Land. There was a fair amount of hate in there, too. A letter of instruction written by Pope Urban in 1095 states that "a barbaric fury has deplorably afflicted and laid waste the churches of God."[23] Descriptions of Muslims as barbarians, Antichrists, "a despised and base race, which worships demons," and other such terms of hatred pepper contemporaneous accounts of the Crusades.[24] The crusaders hated their enemy, deeply. To these men, Muslims were more than an enemy; they were a threat and a danger. From an Augustinian point of view, the enemy was a group of sinners. "Barbaric fury," worshipping demons, and that sort of thing was to be opposed; it was, Augustine might have thought, a righteous hatred born from a love of God. The Christians who chronicled the Crusades might have believed that to put people they thought were in the business of spreading other religions—sinners, in their view—to death would also have been an act of *uti* love. The crusaders and their chroniclers were certainly not saints, even if the man whose understanding of love they followed was.

Pope Urban used Augustinian notions of love to motivate the first crusaders, to firm up his position as the one true pope, and to try to reunite the Orthodox and Christian churches. In those first two things, he was pretty successful. In the last, not so much. The two churches continued to drift off in their own directions, not helped at all by those members of the Fourth Crusade who decided to invade the Orthodox Church's homelands. But the need of so many in Europe to express their *uti* love—to do something in the name of God to help them on their path toward blessings—went far beyond what Urban could possibly have dreamed. Of course, offering them time off in purgatory and a share of the loot for their efforts didn't hurt, either. It turns out that expressing *uti* love could be really—well, useful.

From Blood Sacrifice to Loving Grace

"God is love," you might hear Christians say. When they do, they might be unwittingly quoting Saint Augustine. It's a simplified version of Augustine's description of love as an affectionate attachment based on valuing, caring for, and wanting to unite with God. Modern Christian ideas about God also hold that this love flows both ways, claiming that God and Jesus love you back just as thoroughly as you love them. That's how robust the emotional regime Augustine created was and still is. He developed an understanding of Christian love that remains at the religion's core when that religion is at its best. The notion that to love your neighbor is to use your feelings in the service of God and Jesus, and that feelings are not good or bad in and of themselves but rather are only of value in how they are used, still echoes down the aisles of many a Christian church. It is fair to say that the world would look very different without Augustine's ideas about love. Were it not for his influence, Christianity might still be based on the blood sacrifice of a messiah rather than on an act of divine charity and grace by a

god. It's hard to say how many of the 2.4 billion people who call themselves Christians would be Christian without it.

As for the more recent past, when I say that Urban's interpretation
of love shaped history beyond his dreams, I don't just mean in terms of
the Crusades. The aftermath of that series of conflicts was three centuries of brutal war between European Christendom and the Islamic
Middle East, until one of them emerged as a great unified power in
1396, during the Battle of Nicopolis, with a massacre that introduced a
new Islamic empire—the Ottomans. That's where we head next, to
the moment when the Ottoman Empire changed everything. Albeit
by accident.

Five

What the Ottomans Feared

Outside the walls of the great city of Constantinople, on the evening of May 28, 1453 (or 857, according to the Islamic Hijri calendar), a man wandered through the ranks of the Ottoman army. His hood up, he moved from one group of soldiers to the next, battalion to battalion, unit to unit. This group of warriors had been laying siege to the city for six weeks, pounding the walls with cannon fire. And not just any cannon fire. One of these cannons was the eight-meter-long Orban bombard, a battery so uncomfortably massive and powerful that it could only fire seven times a day. Those six weeks had been grueling, stressful, and not always entirely successful. Morale had started to wane. But their sultan, Mehmed II, was very good at building confidence. On this day, his skill was especially important because on the next, they were attacking Constantinople in a final push. The day of atonement and prayer that the sultan had called for had focused their minds. It had helped them remember why they were there and how grand a prize the conquest of Constantinople would be.

If the truth be known, it wouldn't be all that big a prize on paper. At least not from an outsider's point of view. The sparkle of the once great city of Constantinople, jewel of the Byzantine Empire, had begun to dim ever since the Crusades started. It had become little more than an excuse for war; a stop for weary knights on their way to greater glories.

On one occasion, that stop had become the sacking and conquering of Constantinople itself—during the Fourth Crusade, in 1204. Things had only gotten worse since the arrival of the Ottoman and Seljuk Turks on the scene. The Byzantine Empire had been chipped away, its capital city's population reduced, and its power and influence diminished.

But Constantinople held deep symbolic significance for the Ottoman soldiers who were banging on its gates. This wasn't the first time Islamic forces had tried to attack. The first attempt came in 678, when Mu'awiyah I, caliph of the Umayyad dynasty, strove to take the city, losing soldiers who had been close to the Prophet Muhammad in the process. Even Mehmed II's father, Sultan Murad II, had tried to overpower the city walls, laying siege in 1421. On that occasion, he'd had to abandon his efforts when trouble erupted elsewhere in his empire. But despite their opponents' luck, there was a belief—a thought, a sense, a feeling—among the Ottomans that Constantinople meant something special to Allah, that God himself wanted it to fall to Islam. To the Ottomans, the city was *der saadet,* the "gate to prosperity." To Murad's son, it was unfinished business.

As the wanderer passed through the camp, he heard the holy men among them reciting the names of those companions of the Prophet Muhammad who had died in 678. He also discovered that a *hadith* had become increasingly popular on the lips of those he encountered. It went like this:

Muhammad asked, "have you heard of a city surrounded on one side by water, on the other by land?" They said "yes." Muhammad continued by prophesying: "the Final Day will not occur until 70,000 sons of Isaac arrive to conquer it. They will enter without shooting an arrow or firing a weapon. The first time they say 'there is no God but God and God is most great' [*la illaha illa la wa allahu akbar*] the sea side of the city will fall to them. The second time they say the same phrase the land side will fall. When they say it for the third time, their way will be cleared, they will enter the city, and take its booty."[1]

This *hadith* was likely written specifically for the attack on Constantinople. But that doesn't matter. It has the right spirit to convey what focused the soldiers' minds that night. For them, as for the sultan, the taking of Constantinople would not only demonstrate the power of the Ottoman Empire under Mehmed II. It would also be an act performed in the service of Allah, to gain his love and his mercy. They were compelled by a deep emotional drive that had its origins 843 years earlier, at the heart of the Koran itself. Their drive was fear, but not a bad fear. A fear that could inspire great things.

What Is Fear?

According to modern science, fear arises when various areas of the brain—most important, a group of cells at the base of the brain called the amygdala—perceive a threat. The amygdala is part of what some psychologists used to call the "lizard brain"—an ancient part of the brain that reacts without the need for thinking. These days, few still hold to that idea, because the brain seems to have evolved in strange and complicated ways. But the point is, the amygdala reacts quickly. Often so quickly you don't even know it's reacting until after it's made you do something like jump or run away.

Fear researchers have described the emotion as "a motivational state aroused by specific stimuli that give rise to defensive behavior or escape."[2] That's the urge to fight, flee, or freeze. But that's not all there is to fear. Fighting, fleeing, and freezing don't cover the whole range of behaviors that we associate with fear. Some people might feel frightened when they see a mouse. They might experience the urge to fight, flee, freeze, or climb on a chair. But others live with a permanent dread of the mouse around the corner, and this dread can range from an active dislike to a full-blown phobia. The latter is often the result of some sort of trauma. It can lead to what is known as hypervigilance. Like fear, hypervigilance can both keep you alive and—if it's out of control—make your life unbearable.

The kinds of fear associated with phobias, like panic, come from the locus coeruleus. That's the bit of the brain stem that produces a lot of noradrenaline, the same neurochemical that creates affection when mixed with dopamine. That's not to say that love and panic are linked—well, not always. Although both can cause trembling, butterflies in the stomach, and the desperate urge to run away, the noradrenaline release associated with fear doesn't mix with dopamine.

Like disgust and love, fear may have some evolutionary neuro-chemical origin, but its experience is also cultural. There are clearly some things that nearly everyone is frightened of, such as falling off the edge of a cliff and becoming fatally ill. But fear is quite hard to pin down in universal terms. For every arachnophobe, there's someone (such as me) who loves spiders and even keeps giant tarantulas as pets (which I do not). Even among people with supposedly universal fears, there's someone happy to free-climb a rock face.

There are fears that we learn through culture, upbringing, and education. Not just obviously taught things such as the fear of hell and damnation and the threat of aliens but also some things you assume are universal, such as basic health and safety. Most of us were taught to be frightened of moving vehicles as children. As adults, we might find it nearly impossible to cross a busy road without taking precautions. But there are people from cultures around the world who happily cross with less caution, people who don't share our deep fear of getting harmed by traffic. Anyone who has ever traveled to India or driven in Italy knows what I mean. Maybe the fear we learn is something quite different from innate fear. Perhaps those non-innate fears are, like love, an emotion complex: an emotion built from a complex group of dreads, worries, panics, states of hypervigilance, and genuine concerns that attach themselves to whatever it is that makes us fearful.

It's worth pointing out that fear isn't always a bad thing. Scientifically speaking, any feeling that keeps you from being harmed isn't just good; it's also an essential part of evolution. People who are afraid that there might be a bear hiding behind a bush tend not to get eaten by bears. Nevertheless, modern science likes to define whether an emotion is positive or negative by how it makes us feel. That's actually a relatively new idea—the birth of which I'll get to later in this book. The idea that fear is bad would have been alien to the Ottomans and most of the people they encountered. Whether an emotion was considered good or bad came down to how it was used, the purpose to which it was directed. We've come across this idea before with the Greeks' notions of *eros* and *ataraxia*. We've also touched on the good fears that existed and still exist in other religions, even though we haven't yet focused on them. The ancient Hebrews were fearful that they might cause God to be disgusted with them. To Augustine, the fear of God can lead you to God and heaven. The fear the Ottomans felt was just this sort of positive fear. It was a type of fear that drove them to remain steadfast in their faith and accomplish great things. It was the positive fear of Allah, and it can be traced right back to the founder of their religion, Muhammad.

The Mountain That Moved Muhammad

To understand this godly fear, we need to go back to 610, when a forty-year-old man by the name of Abu al-Qasim Muhammad ibn 'Abd Allah ibn 'Abd al-Muttalib ibn Hashim—better known simply as Muhammad—began climbing to the top of Mount Hira, in the region now known as Saudi Arabia. Moments earlier, he, like Saint Augustine before him, had heard a voice telling him to read. But this wasn't the voice of a child. This was the voice of something otherworldly, the pronouncement of a being of extraordinary power. The entity had appeared before him, filling his field of vision at every turn. Initially, Muhammad refused to speak, so the creature ordered him to read a second time. Skeptical of the apparition's origin, he refused again. He was terrified that it might be a jinni or some other preternatural entity sent by the devil to trick him. Once more, the voice made the order, and once more, Muhammad refused. The creature then embraced him until he could stand it no longer. Muhammad found himself losing control of his own voice as the first words of the Koran burst, unbidden, from his mouth.

Read in the name of your Lord who created; created Man from clots of blood.
Read! Your Lord is the Most Bountiful One, who by the pen taught Man what he did not know.[3]

Muhammad was terrified. Overwhelmed by the horrific realization that he had, indeed, been possessed by a demon, or jinni, he rushed up the mountain. He was ready to release himself from whatever had taken control of his flesh by casting his body onto the ground below. His life flashed before his eyes.

Muhammad grew up in a melting pot of beliefs. His tribe adhered to a variation of the ancient Semitic religions popular in the region, which placed their faith in a range of gods, including a fertility god called the Goddess, or al-Lat. But most important of them all was a supreme Father God called the God, or al-Lah. These weren't the only gods Muhammad knew. The sacred shrine—known as the Kaaba—at the center of his hometown, Mecca, was fashioned from a beautiful piece of jet-black meteorite. Of course, the citizens had no idea it was a meteorite at the time. It was just a bit of stone so otherworldly and striking in its appearance that it seemed a suitable home for deities. It acted as a temple to the many gods revered by the clans who lived in the area. Then, as now, tradition dictated that a person was to walk around the stone seven times before approaching for worship.

Muhammad was a trader whose domain stretched across the Arabian Peninsula. He was greatly respected for his honesty and his relentless hard work. It seems likely that he absorbed the stories and beliefs of the religions from outside his region as keenly as he did the local ones, leading to something of a crisis of faith. This is why, when entering middle age, he began to spend weeks at a time in a cave in the mountains, indulging in the spiritual practice of *tahannuth*. Once alone, he would focus his mind on prayer, breaking away only to feed the poor. Muhammad had started to wonder if the head god, al-Lah, was the same one true God he kept hearing about from Christian and Jewish merchants.

It was during one of these rituals that he is said to have heard the terrible voice that took over his own. Of course, he'd had visions before—that was part of the ritual—but nothing so vivid, so powerful. He was convinced that he had been possessed by something malevolent. Muhammad felt he had no choice but to climb the mountain and throw himself over the edge before the evil overcame him, potentially harming the people he loved.

The story goes that as he climbed, the vision returned, even more terrifying than before. This time, it introduced itself: "O Muhammad! Thou art the Apostle of God and I am Gabriel!"[4] The apparition left Muhammad frozen in terror. He didn't move a muscle until the search party sent out by his first wife—Khadija—found him sometime later and helped him back down the mountain.

Muhammad remained confused and terrified for some time afterward. But the apparitions kept coming, and whenever they did, Muhammad turned to his wife for comfort, begging her to "cover" him as he fell, shivering, into her arms. Muhammad was experiencing the raw, extraordinary power of the one true God: Allah. He knew that this power was to be feared, and before long, he discovered that his fear could be used as a force for good. He had become a prophet—a mouthpiece—for God. Amid those terrifying visions, Muhammad was learning something. He was experiencing firsthand how passions might be used to control people's hearts. The words he spoke to others would come to form the Koran, the religious text at the center of the Islamic faith.

The Beautiful Fear of God

The text of the Koran itself is beautiful. After it was compiled into its present form in 650 by Abu Bakr, the book's content took on an emotional aspect that compels just about anyone reading it to treat it with respect. Ask any adherent of the Islamic faith, and they will tell you that the original Arabic text is beautiful—perfect, even. Some believers claim that the Arabic contained inside is so flawless that it had to be divinely inspired. There is not one phrase out of place, not one word poorly chosen, not one grammatical misstep. It should be read with at least as much emotional weight as a performance of *Hamlet*, and probably even more. In fact, the adulation that actors shower on the words of the great Bard is nothing compared to what followers of the Koran devote to this most sacred text. Shakespeare is to be performed, not just read. The Koran, whether read out loud or silently, should be accompanied by a similar performance.

One *hadith* reads:

This Koran was revealed with sorrow, so when you recite it, then weep. If you cannot weep then pretend to weep, and make your voice melodious in reciting it. Whoever does not make his voice melodious, he is not one of us.[5]

Readers of the Koran are supposed to take their time, allowing several days to complete the entire text if necessary. They should weep when they are supposed to, speak in beautiful tones when the text is conveying something beautiful, raise their voices and soften them where appropriate. Readers should contemplate the words and let their bodies act accordingly. People in the faith believe these to be the words of Allah himself, not just those of a gifted playwright or an apostle.

The Koran is a book that evokes emotions purposefully in an attempt to change readers' lives and bring them closer to Allah. It's a work designed to be responded to with weeping and loud proclamations. Certain *surahs* (chapters) and revelations should make readers react in visceral, emotionally moving ways. Like most religious texts, it exerts an emotional influence, designed both to transform people into Muslims and to reinforce their faith after they have converted. The key to that transformation is a type of fear, but not fear as you might understand it.

The Koran is a work in two parts.[6] The earlier, Meccan *surahs* are where most of the really emotional content is found. These are the *surahs* recorded before Muhammad moved to Medina, in 622, to avoid being killed by people who weren't happy about his going around converting everybody to his new religion. Perhaps because he didn't have the freedom to preach openly, which he'd later find in Medina, or perhaps because they had to be memorable, these early *surahs* are shorter, more rhythmic, and focus a great deal more on the beliefs of the region's other religions. It's in these *surahs* where you are most likely to find mentions of Adam and previous prophets such as Moses, Abraham, and Jesus. These chapters emphasize the importance of morality, the dangers of worshipping the wrong god or gods, ways to get to heaven, and ways to avoid hell. They are much more severe and somber than the *surahs* Muhammed composed later in life, after conquering Medina. It's not that the latter *surahs* aren't emotional—far from it. It's just that Muhammad's understanding of emotions, particularly fear, is more clearly expressed in those shorter chapters, so that's where we'll begin.

In the Meccan *surahs*, Muhammad exhorts people to believe with intensity and focus. That's not surprising: these were the *surahs* recited as he began to build his following and spread his word. The *surahs* had to be snappy, sharp, and easy to learn by heart, because a strong oral tradition was critical to the early spread of the faith. It's believed that some of the *surahs* were recorded on whatever Muhammad's followers could find at the time — cloth, bits of bone, spare pieces of leather, and so on. Most of these shorter *surahs* are now found at the end of the Koran. Current versions arrange the *surahs*, for the most part, by length, with the longest appearing first. To read what the younger Muhammad said to his followers in those early years, you need to go to the end of the book. It is here that you also find some soft echoes of the ideas of Saint Augustine.

There's no evidence at all that Muhammad knew of Augustine's work, but he could well have. As I noted above, Muhammed was a trader with a keen interest in a variety of cultures and traditions. It's entirely likely that he picked up a great deal of knowledge about both Christianity and Judaism from other traders and from the holy men he encountered in his work. Muhammad built up an almost encyclopedic knowledge of these faiths, albeit filtered through his own culture. Such familiarity with the ideas of Christianity might suggest that he'd at least heard of Augustine, given the latter's prolific output and fame. But regardless of whether Muhammad knew about Augustine specifically, the two shared some strikingly similar ideas.

Like much of Augustine's work, the Koran emphasizes the difference between godly love and self-love — and the virtue of the former over the latter. It reminds its readers repeatedly that Allah gave humans both the ability to think and the ability to feel. The message is that if you love Allah, Allah will love you back. If you don't, well, then there is literally hell to pay. Also, like Augustine's, the Koran's conception of love isn't just about charity and loving others for personal gain but about actively desiring the love of Allah. "People's wealth you seek to increase by usury will not increase in the sight of God; but the alms you give for the love of God shall be repaid many times over."[7] In plainer terms, loving your neighbors ought to be done selflessly in the name of Allah.

But love is not given freely by Allah; it is something you must earn. Similarly, mercy only comes at the behest of God. To gain love and mercy, one must act as a Muslim ought to, as directed by the Koran. This is what makes the Koran different from the works of Augustine. A type of fear, albeit one based, like Augustine's, on love, was important to Muhammad's new religion.

It's pretty much always been the case, from the Greeks to modern science, that *fear* has been something of a catchall term that covers a wide but related range of feelings. For example, consider the difference between disliking mice because you find them generally unpleasant and disliking them so much that you have full-blown, hypervigilance-causing musophobia. But it's not just a matter of degree. There are many subtle distinctions among definitions of fear to be found across the scientific literature and within philosophical and religious texts. The Koran mentions ten types of fear, including *khawf* (the fear of a danger you see coming and should prepare for), *khashiyah* (the fear of the harm something can do to you), and *taqwa* (the type of fear that leads you to take precautions, such as locking your house or wearing a mask during a pandemic). These can all be experienced in the service of Allah, but they can also be experienced for selfish reasons. The Koran quotes Allah as saying, "It is Satan that causes his followers to be feared. But have no fear of them, and fear Me, if you are true believers."[8] Augustine distinguished between self-love and love focused on God. This love of God could also generate a fear of sin—an idea that takes us right back to the Hebrews' fear of causing God to be disgusted. The Koran does something similar. It distinguishes between ordinary, run-of-the-mill fears of earthly terrors and the right type of fear—the fear of God, or, rather, the fear of letting God down.

Although most translations of the Koran use the term *God-fearing*, the Arabic words don't mean that. Not quite. They refer to one of those emotions that's difficult to translate. The original Arabic words for "fear of God," *alkhawf min Allah*,[9] are more accurately understood to mean something like "God consciousness" or "God-protecting." It's a fear of losing sight of the tremendous power of Allah. Of developing an aversion to God or not protecting the faith from those who have an aversion to it. Of losing control and not behaving as a Muslim ought to. To put everything in your life second to Allah—your property, your family, even your own life—in order to avoid feelings of loss. That loss, or *khusr*, is also an essential part of the Koranic emotional makeup. One of the last and shortest *surahs* in the Koran says:

I swear by the declining day that perdition [or loss] shall be the lot of Man, except for those who believe and do good works; who exhort each other to justice and fortitude.[10]

According to the Koran, not behaving as a Muslim ought to is tantamount to losing, or perhaps simply never receiving, Allah's love and mercy. It's a loss that is a path to perdition, to hell and damnation.

Fear of God helps believers remain mindful and protective of their focus on Allah. It enables the faithful to go against their natural passions, to keep their emotions under control, and do what is right by Allah rather than act out of pure self-interest. That focus will lead you to prosper, to win in battle, and to gain your rewards in heaven.[11] The message is clear—if you genuinely fear letting Allah down, you will accept what the Koran has to say, keep your emotions in check, and gain your rewards.

Another important element of the emotional aspect of the Koran is the role of the heart. There are four often interchangeable words for "heart" in the Koran. *Qalb* and *fu'ad* literally mean "heart." *Lubb* means "the internal heart" or "mind." Finally, *sadr* means "chest" and/or "breast." Collectively they appear 208 times in the text.[12] The Koran places a person's emotional center in the heart. It also puts the seat of *aql*—meaning "reason" or "understanding"—in the heart.[13] The head, or the *ra's*, is just a head. This means that, in the context of early Islam, thinking and feeling weren't understood as two separate things. According to the Koran, nonbelievers' hearts are hardened not only because they don't understand the message of Islam but also because they don't *feel* it. The idea of the heart as the source of our emotions comes up a lot in ancient literature. The fact that we still say "I feel it in my heart" is a testament to the depth with which this concept penetrated so many spiritual and cultural traditions. Perhaps, because it's an essential part of the way emotions were understood in Ottoman Islam, it's time I addressed it by getting a bit medical. To do so, we need to move forward to one of Islam's greatest intellectuals—a man by the name of Abu 'Ali al-Husayn ibn 'Abdillah ibn al-Hasan ibn 'Ali ibn Sina, or Ibn Sina for short. He was known to the West as Avicenna.

With Great Humor

Ibn Sina was born sometime around 980 near Bukhara, in what is now Uzbekistan. The son of a government official, Ibn Sina had memorized the entire Koran by the time he was ten years old. He mastered Indian arithmetic, Islamic law, and a great deal of ancient philosophy before his teens. Shortly thereafter, he began a relationship with the works of Aristotle. He reread *Metaphysics* until he knew it word for word. More important, he learned to understand it intimately with the help of commentary written by another great Islamic thinker, al-Farabi. At sixteen he decided to study medicine, rumor has it, because it was easy. His medical skill became widely known, and he wrote two books on the subject—*Kitab al-Shifa'*, or *Book of Healing*, and *Al-Qanun fi'l-tibb*, or *The Canon of Medicine*.

One of the biggest influences on Ibn Sina, and indeed on every physician who lived anywhere from the western borders of India to the shores of Ireland, was Galen. Medically speaking, Galen may well be the single most influential thinker in the history of Western civilization. His ideas dominated medical practice for well over a millennium. Galen mattered. He mattered a lot. Unlike lofty ideas of theology and philosophy that meant nothing to the average farmer, Galen's views on healing affected everyone, king and peasant alike, though peasants were more likely to treat themselves using his principles than employ the services of a physician. As is too often the case, the fact that he was wrong about pretty much everything didn't matter.

Scholars in the great Islamic empires that arose soon after Muhammad's death translated and preserved Galen's words, sometimes embellishing them and adapting them. In the twelfth century, his teachings made their way to Europe and became the backbone of most of the Western and Middle Eastern world's approach to medicine until germ theory finally supplanted them once and for all in the 1800s.

Galen borrowed many of his ideas from the Greek physician Hippocrates—the man from whom we got the Hippocratic oath. Hippocrates believed that humans were composed of four humors. Each of these humors had a particular temperature and consistency. The warm and wet humor around the liver was *haima*, or blood. Cold and damp *melaina chole*, or black bile, gathered in the spleen. The gallbladder made the warm and dry humor *xanthe chole*, or yellow bile. Finally, the brain produced cold and wet *phlegma*. (*Phlegma*, by the way, didn't mean "phlegm," as in "spit." It was any clear substance found in the body.) Galen believed that for the body to operate efficiently, all four humors had to be perfectly balanced. Striving to maintain a healthy diet, get adequate sleep, and exercise regularly was at the center of Galen's medical practice. (I did say he was wrong about "pretty much" everything.)

The important thing with regard to our present topic of emotion is that each humor was associated with a mood. Many languages use versions of these words to describe feelings in the present day. Blood was linked to being sanguine—chatty, active, and extroverted. Black bile, or *melaina chole*, is the source of the English word *melancholy*, and that just about sums up its effect. Too much yellow bile made you choleric—easy to anger, tetchy, and stressed out. An excess of phlegm made you lazy, lethargic, and lifeless—phlegmatic. But for Galen, that's about as far as it went when it came to the humors. The only treatment for them was to behave the right way and control them.[14] But just as important, to both Galen and to Ibn Sina, was breath.

Galen argued that the soul animated the body using *pneuma*, or air. There were three types of *pneuma*: *Pneuma physicon* (or natural spirit) resided in the liver and helped regulate our vegetative selves, governing things such as nutrition, metabolism, and reproduction. *Pnuema psychicon* (or animal spirit) was found in the brain and controlled the "animal" aspects of our lives, such as sensory perception and movement. *Pneuma zoticon* (or vital spirit) was located in the heart and controlled body temperature and the flow of blood and humors.[15] Ibn Sina took this notion and ran with it. He believed that breath originates in the heart as the breath of life—the very air Allah breathed into Adam at creation. This soul-breath then filters through the body, interacting with the humors and allowing the organs to do their job. In his view, the purpose of humors is to build and interact with parts of the body, sometimes causing changes in behavior, such as the expression of unruly emotions. When the flow of breath is restricted or altered, this could unbalance the humors and cause illness, including diseases of the passions.

Ibn Sina didn't write a systematic analysis of emotions. But he did write about ways in which the passions could be detected using a combination of a patient's breathing and his or her pulse. Deep breaths and a pulse that was "large, rises high," and was "brisk" suggested anger. A sudden, quick, irregular pulse with rapid breathing indicated fear.

(The speed with which these symptoms appeared correlated with the degree of the terror.) Grief was accompanied by an almost undetectable pulse and weak breath. Finally, gradual breath and a slow, frequent, but relatively powerful pulse meant that the patient was experiencing joy or delight.[16] The joyful patient would also be warm, as hot, wet blood became the dominant humor in his body.

Ibn Sina placed feelings firmly inside the body. To him, the passions were a medical issue, and cures for unruly emotions no longer had to involve religious penance, prayer, or exorcism. You could find the right treatment by balancing the humors. It also meant that monitoring the source of emotions—the heart—could help control your feelings and keep you on the path Allah set for you. This idea meshed with the Islamic teaching that your emotions should match that initial divine breath of thought and feeling, which slows the pulse, softens the heart, and leads to Islam. This, at least, is likely to have been what the Ottoman army believed as they prepared for the attack on Constantinople on the evening of May 28, 1453.

The Community of War

Let's go back to the siege outside Constantinople, where, according to the sixteenth-century Ottoman historian Neşri, Sultan Mehmed II was making a speech.

> These tribulations are for God's sake. The sword of Islam is in our hands. If we had not chosen to endure these tribulations, we would not be worthy to be called gâzîs [warriors]. We would be ashamed to stand in God's presence on the Day of Resurrection.[17]

This might have been one of many speeches Mehmed made. On the days leading up to the final push, he kept himself busy. He rode around the camp on horseback, speaking to his men and his commanders. He organized assaults and made himself part of the greater war effort, part of the community, rather than some distant figure in a tent, miles away.

Emotional Communities

The Ottoman army was made up of people from a variety of communities called a *millet*, or a *taife*. Sometimes these were separated by ethnicity, sometimes by faith—not everyone fighting for the Ottomans was Muslim, though the vast majority were. Despite these separate communities, during battle, there was an overarching emotional community that bonded all the soldiers together. Fundamental to this emotional community were ideas such as *rıza ve sukran*, a consent or gratitude given to those who were pulling their weight, doing what they were required to do, and refraining from being offensive to the group. The latter was especially important. The Turkish phrase relating to not being offensive—*kendi halinde olmak*—literally means "be on its own," which seems a little odd at first glance. But *kendi halinde olmak* didn't mean being on your own in terms of separating yourself from the group so as to avoid upsetting people or anything like that. According to historian Nil Tekgül, it meant trying to be "inoffensive with an emphasis on the harmlessness"[18]—not causing trouble, showing off, or being a pain in the neck. Just getting on with your job.

Community is, of course, an important part of the Koran. *Ummah*—a type of community commanded by Allah himself—is supposed to bind followers of Islam together:

> Apostles! Eat of that which is wholesome and do good works: I have knowledge of all your actions. Your community [*ummah*] is but one community [*ummah*], and I am your only Lord: therefore fear Me.[19]

The *ummah* shared within and among the *taife* was part of its emotional makeup and the emotional makeup of the Ottoman army itself.

As sultan, Mehmed was responsible for reinforcing those communal bonds. He had to show his consent and gratitude to his men for what they were about to do, and he had to demonstrate that he, too, feared Allah and wanted to do right by him. Doing so was crucial for ensuring that his troops would remain a tight, single-minded unit: one great *ummah* built from individual *taifes*. He, like they, would be ready to join the collective fight, to show gratitude to one another, to use their fear of Allah to soften and focus the breath of God in their hearts so they might take Constantinople—to do as Allah commanded in return for his mercy, love, and glory. It was up to the sultan to address all of them, to help them prosper.

It is likely that Mehmed believed his troops had been put under his protection by Allah. He was charged with fulfilling codes of protection, or *siyanet*, and codes of compassion, affection, and tender kindness, or *merhamet*. It was up to the sultan to help the hearts of his men prosper (*müreffhü'l-bal*) and remain in a "still-state" (*asude-hal*). It was up to him to keep his men as happy as possible.[20] Even during wartime, this mattered. And that's probably why the sultan laughed and chatted with the boys and showed himself to be one of them. He was displaying compassion for his subjects as their ruler and as part of their emotional community.

Suffice it to say that it worked. Bonded as a unit, the Ottoman army successfully conquered Constantinople the following day. The Byzantine walls crumbled, the army fell, the empire collapsed. The last vestiges of the Roman Empire, once so dominant in the Western world, were snuffed out—except the Catholic Church, of course, but that's not important right now. What is important is that Sultan Mehmed II had achieved what had eluded so many. Istanbul, as Constantinople would come to be known, became his most prized possession, his new capital. It was to be brought back to its former glory and made beautiful again. The Ottomans need not fear Allah—the spark of life in their hearts had driven them to victory in his name. Allah need take no offense.

It's difficult to overstate how great an impact the Ottoman invasion of Constantinople had on history. To begin with, modern Turkey still clings to many of the myths surrounding the fall of the city. Without the fear that drove the search for Allah's mercy beating in the hearts of Mehmed II's men, there wouldn't even be a Turkey. The Ottoman invasion of Constantinople wasn't some tactical decision. It was fueled by an emotional drive in Mehmed and his men, a desire to finish what their fathers had started in the name of God. It was made possible by the emotional communities of his men.

The fall of Constantinople had enormous consequences that Mehmed himself couldn't have foreseen. It gave the Ottomans control over access to the Silk Road, and, of course, Mehmed took the opportunity to levy backbreaking taxes on the merchants who traversed it. For more than two millennia, people had traveled from China to Europe and back again along these pathways. It had been an essential part of the European economy. Suddenly, traders could no longer afford these journeys. The repercussions of this would redraw the map, changing Europe and the world forever.

The evaporation of commerce along the Silk Road is, arguably, responsible for creating the part of the world we now call Europe. The once blurred edges of Christendom had become sharper, its boundary more obvious. These new Europeans had to find a way around those sharp edges by taking daring and undreamed-of risks to travel to the East. What happened when they did, and the emotions it unleashed, defined the early modern age.

Six

Abominable Witch Crazes

Imagine, for a moment, that you're a woman of advancing years living in seventeenth-century Europe. You are lonely. You don't have any children, and your husband died some time ago. The only way you can keep yourself alive is to rely on the kindness of the people in your village and occasionally make medicines for them. You're lucky because the villagers think of you as a "wise woman"—a bit of an oddball who dabbles in some white magic, but generally harmless. Then one day in the mid-1610s, a cow belonging to someone who had been rude to you sickens and dies. In a flash, you are no longer a wise woman but a witch.[1]

For months, even years, people regularly call you a witch and are nasty to you. Eventually, you break. You scream, shout, curse, and swear at the villagers, becoming enraged to the point of vengefulness. In your rage, you are ready to do anything to enact that vengeance, even turn your back on God. At that point a dog appears and speaks to you. You are a little scared: after all, it's a talking dog. That fear only intensifies when the dog tells you he is the devil himself. The devil dog asks you to calm down because he loves you "much too well to hurt or fright" you. You are lonely, bitter, and hurt, and here is a talking dog offering you love. It actually feels quite nice to be loved after all these years. The devil dog then offers you the power to take revenge on all who have wronged you, and all he wants in return is your body and soul. Giving up your immortal soul might be too much to ask for a little bloody vengeance, so the devil adds an incentive. If you refuse, he will "tear [your] body in a thousand pieces." What would you do?

According to a play first performed around 1621 and supposedly based on "a known true story," this was precisely the dilemma that faced Elizabeth Sawyer. In the play, "Mother" Sawyer gets so sick of being treated poorly that she jumps at the deal, wreaking madness, suicide, and murder on her neighbors before being caught and hanged.[2] However tall this tale might seem, it's not that unusual when compared to other stories about how people become witches.

The witch crazes of the sixteenth and seventeenth centuries are some of the worst examples of violence against women in all of history. Best estimates of the number killed between the years 1560 and 1630 hover around fifty thousand, though this is debated.[3] Some say fewer; some say more. Some have claimed it's in the millions, though that's unlikely.

One of the biggest witch crazes happened in Trier, an ancient Roman town now in Germany, over the twelve long years between 1581 and 1593. As many as one thousand people accused of being witches were killed there after a particularly nasty Catholic archbishop—Johann von Schönenberg—decided to cleanse the town of Protestants, Jews, and witches. He drummed up widespread fear, aided by secular prosecutors who hoped to profit from the widespread suffering—the trials weren't cheap, and the Church was rich. Soon enough, people accused of witchcraft were dragged out of their homes, tortured, and burned alive. Often, the process would lead to further accusations, snowballing into mass hysteria.

You may now be asking yourself a question that's plagued historians for decades: How did this happen? What conditions could have possibly created a situation in which thousands of people—mostly women—were killed for a crime they almost certainly didn't commit? The answer, or at least a big part of it, has to do with emotion.

Of the many passions people felt toward witches in the period, two were particularly important—and these are feelings that we've come across before, at least versions of them. These two feelings are the keys to explaining why so many people were killed. The first is fear, and the second is a particular type of disgust—*abomination*.

The way the people of Europe understood fear and abomination at the time of the witch crazes has its roots in a much older emotional regime—one that combined Christian teachings with Greek thought. This wasn't the work of Saint Augustine but that of a man who developed Augustine's ideas into an intellectual force that dominated Christendom for the following four hundred years: Saint Thomas Aquinas.

The First Book About Emotion

The twelfth century has been described in many ways—the twelfth-century renaissance and the twelfth-century crisis among them. What is clear is that it was a century of profound change in Europe, and much of that change came back with the crusaders. An increase in bureaucracy is one significant, if somewhat prosaic, example. The number of lists of goods, inventories, and the like increased dramatically in the twelfth century. There was also a sudden influx of Latin translations of ancient Greek literature, including the works of Aristotle.

The impact of Aristotle's twelfth-century reintroduction to the intellectual lives of Europeans was immense. Ideas based on Aristotle's works dominated education and debate for the ensuing five hundred years. Many historians once thought that it sent us down a very dark path. They argued that the Church's ideas, filtered through Aristotelian logic, put a stranglehold on progress for half a millennium, leading to the so-called Dark Ages. Of course, it's not that simple. There were many great intellectuals and inventors in the period. Everything from the horse collar, which revolutionized farming, to the legal marvel that was the Magna Carta appeared around this time. Some excellent philosophers were writing profound and influential works then, too. One was a Dominican friar named Thomas Aquinas.

By the thirteenth century, the profound changes brought about during the preceding hundred years were taking hold. It's at that time that Thomas Aquinas wrote, among other things, an unfinished work called the *Summa Theologiae*. The *Summa* is a colossal book, both in length and intellectual scope. In it, ideas are revealed point by point. First Aquinas makes a statement. Then he offers up objections to his own account. Next he analyzes his objections before finally drawing a conclusion. That conclusion becomes the next statement, and the whole process starts again. It's easy to get lost in this work, and it's incredibly easy to take Aquinas out of context by quoting the wrong parts of the argument.

Most important for us is one specific part of the *Summa*—sections 22–48 of the *Prima Secundae*, or first part of the second part. These chapters almost certainly comprise the first dedicated work about emotion ever written. Aristotle had systematically covered emotion in *Rhetoric*, but not for its own sake; in *Rhetoric*, Aristotle's descriptions of *pathē* were part of a work about how to debate. Aquinas actually dedicated an entire section of his book to emotions. I say "emotions," but technically he was writing about the category of feelings known as passions, or *passiones animae* (passions of the soul). These were feelings that began in the body and influenced the mind, not unlike *pathē*. Aquinas also identified another category of feelings, *affects*, or *affections*, that travel in the opposite direction: the mind thinks about something for a bit, perhaps even for a long time, then makes the body feel the appropriate response. To put it simply, passions are the feelings that Plato and the Stoics so desperately wanted people to control. They were the cause of Augustinian self-love and the sin that followed. Affects were the controlled, mindful, good feelings—the ones that, according to Plato, lead you to *eros* or, according to Augustine, begin with *caritas*. Some people, such as the Stoics, believed it was possible to turn one into the other.

Aquinas placed the individual passions onto a list that borrowed heavily from Plato and Aristotle. This menu of feelings consisted of the primary passions, which, when mixed, gave rise to secondary passions. He put the primary emotions into two groups of opposing pairs. He borrowed from Plato by calling these groups the concupiscible (or desiring) and the irascible (or angry) passions. But he listed the feelings that made up each category in a way that Plato and Aristotle never did.

The desiring passions were:

* *amor* (love) and *odium* (hate),
* *desiderium* (desire) and *fuga* or *abominatio* (flight or abomination), and
* *delectatio* or *gaudium* (pleasure or joy) and *dolor* or *tristitia* (pain or sorrow).

Like Plato's, these passions were simple, caused by a reaction to something good or evil: a reward or a punishment. For example, *amor* (love) is caused by something good, and *odium* (hate) is caused by something terrible. *Desiderium* (desire) is a wish to move toward something pleasant, and *fuga* or *abominatio* (flight or abomination) is the compulsion to get away from something awful. *Gaudium* (joy) is what you feel when you receive or are close to something beautiful; *tristitia* (sorrow) is what you feel when you are in a dreadful situation.

Aquinas's angry passions were:

* *spes* (hope),
* *desperatio* (despair),

- *audacia* (bravery),
- *timor* (fear), and
- *ira* (anger).

Again, these are the passions that kick in when the going gets tough. Striving for what you desire requires the passion of hope or courage to kick in. Fighting against something evil takes anger, fear, or despair.

Like the concupiscible passions, the irascible passions were paired in opposites. But they were a bit more complicated. Depending on the situation, hope could be the opposite of despair, courage the opposite of fear. At other times, courage was the opposite of despair, and hope was the opposite of fear. The odd one out was anger. Although it was the most irascible of angry passions, it had no opposite.[4]

But Aquinas didn't only categorize the passions as desiring or angry. He also split them up according to when they occur. Some emotions—joy, sorrow, courage, fear, and anger—are about what is happening to you in the moment. So if you win a game, you feel joy at that particular time. If you lose, sorrow. Other passions—desire, flight or abomination, hope, despair—are felt when you know what is likely to happen to you sometime in the future. If you want something but don't have it, you feel desire or hope. Love and hate are outliers because they exist, according to Aquinas, at all points in time.

As I noted above, fear and abomination, as described by Aquinas, were the main ingredients in the European witch crazes. People felt abomination toward women (and men) they thought of as witches, and fear at the time was an unending dread that followed the people of Europe around constantly. Let's unpack these notions, starting with that all-pervasive background emotion, fear, and the reason it hung over people living in the sixteenth and seventeenth centuries.

Terrifying Times

In the previous chapter, we left the Ottomans, who, after their successful invasion of Constantinople in 1453, levied backbreaking taxes on anything coming through their region from the Silk Road—including things Europeans liked a great deal, such as spices, bone china, and silk. Europeans, many of whom had become rich by trading with the East, needed to find a new way to their suppliers. This sent Bartholomeu Dias around the Horn of Africa in 1488—something people thought was impossible because it was so hot that far south that your head might explode. Soon after, in 1492, an experienced sailor from Genoa named Cristoffa Corombo (we call him Christopher Columbus) tried sailing west to reach the Spice Islands. Everyone knew the world was round—they had for thousands of years—so the attempt seemed reasonable enough. The problem was that, unbeknownst to Europeans at the time, an entire continent was in the way. Columbus famously found this out when his ships landed in what is now the Bahamas. To say that the discovery of this new continent stunned the people of Europe would be an understatement. Everything they knew, and everything they believed, was shaken to the core. Well, obviously not everybody felt this way—your average villager probably didn't give a hoot, if he even knew at all. But the educated elites were amazed, confused, and more than a little scared. "What else don't we know?" they wondered.

Twenty-five years later, in 1517, while the world was still trying to get its head around the idea that there was more of the world than they'd realized, an Augustinian monk who had seen all the commotion as a sign from God walked toward the doors of All Saints' Church in Wittenberg, Germany. He carried with him—if the legend is to be believed—a hammer, some nails, and some parchment on which he had written ninety-five theses. Ninety-five criticisms of the Church—not All Saints' Church but the Catholic Church as a whole. This

Augustinian was named Martin Luther, and he intended to enact some tough Augustinian *uti* love.

Luther had started to become frustrated with the Catholic Church, and his biggest frustration was with indulgences. Remember when I briefly mentioned that part of the motivation to go on a Crusade was getting your time in purgatory reduced? By the sixteenth century, you no longer needed to go on a Crusade. Instead, you need only pay the Church to assign monks to pray for you—the more you paid, the greater the number of monks who would pray. Truth be told, these indulgences, as they were known, were instituted because rich noblemen living at the time of the Crusades liked the idea of getting some reduction of their allotted time in purgatory, but they didn't like the idea of marching to the Holy Land and maybe getting killed. Instead, they'd pay for a church full of monks to pray on their behalf. By Luther's time, the buying and selling of these indulgences had become a booming business. Luther thought this wasn't very Christian, so he wrote ninety-five theses explaining why and nailed them to the church door.

It's actually unlikely that he nailed them to the door; that's probably a myth. But he did write them. More important, he spread them far and wide using a new invention—the printing press. His theses were a hit, making them one of the first bestsellers in history. But they also shook Europe to its foundation, causing a widespread rebellion against the Catholic Church that developed into a group of faiths known collectively as Protestantism. The rivalry between Catholics and Protestants was bitter. Religious hatred was used as the pretext for long and bloody wars. Each faction claimed that the other's leader was the Antichrist, the bringer of the end times. Between 1480 and 1700, the leading powers in Europe fought each other 124 times. Some of these conflicts rival the First World War in terms of casualties per capita. Taken as a whole, these various religious wars constitute one of the most devastating conflicts in history. Every European who lived through the period would have witnessed mass bloodshed at some point or other.[5]

On top of that, horrors that had always persisted seemed to be getting worse. Beginning around 1250, the weather across the globe began changing, and by 1550, the worst extremes of the Little Ice Age had set in.[6] Between 1250 and 1650, average temperatures dropped by 3.6 degrees Fahrenheit, and the climate fluctuated wildly, leading to years of famine.

Disease also seemed to be increasingly common. The Plague returned to Europe repeatedly, hitting London no fewer than six times between 1563 and the well-known outbreak of 1665. The Plague was only one of a string of illnesses that struck the people of Europe in the fourteenth century. Measles, smallpox, cholera, and dysentery, exacerbated by the unusual weather and the famines it caused, were also a constant reminder of mortality.

Then there were the big new diseases. These were typically introduced by soldiers returning from military conflicts abroad. While struggling to wrest the town of Baza, in the Iberian Peninsula, from Islamic forces, Spanish soldiers began to notice their comrades coming down with red sores and fever before succumbing to madness and death. This illness, typhus fever, killed more Christian soldiers than the defending Moors did. Typhus soon made its way across Europe, with devastating results. Another new disease, English sweat, caused shivering, giddiness, headaches, severe pains, exhaustion, profuse sweating (as the name suggests), and, most often, death.[7] After first appearing in England in 1485, it engulfed the rest of Europe by 1500.

Whatever the English could do, the French did better. Or should I say worse? While invading parts of Italy in 1495, French troops began to come down with ulcers that developed into a rash on the hands, feet, mouth, and sex organs. With these afflictions came an excruciating agony that would quickly pass. But as soon as patients thought they were getting better, soft, tumorlike growths would appear all over their bodies, followed by severe heart problems and madness. Finally, and mercifully, came death. We now refer to this "French sickness" as syphilis, and the people of the time knew only too well that it was sexually transmitted—a "disease of sin."[8]

By the time of the witch crazes of the sixteenth and seventeenth centuries, all this—the disease, the famine, the increasingly cool nights, and the realization that the ancients, including those who wrote the Bible, knew nothing about the world because they had omitted a massive, great continent from their writings—had put just about everyone on edge. They were scared.

As if all the horrors faced by people living in sixteenth-century Europe weren't bad enough on their own, matters were made worse by the fact that all these unfortunate events mapped neatly onto the book of Revelation, or, as it's known in Catholicism, the Apocalypse of Saint John. Disease? Check. War and rumors of war? Check. Fire? Check. Pestilence? Check. The Antichrist on earth? Check. Everything you thought you knew about the world being turned on its head? Check. It gave people who believed the end of the world was coming—known as millenarians—a louder voice and a veneer of credibility. Thinking you were living in the end times wasn't new—indeed, it still persists to this day. But if there was a time in history when it looked like the earth was convulsing in the throes of biblical Armageddon, this was it. The only things missing were the foot soldiers of the devil himself, and that's where witches and the passion of abomination come in.

Becoming Abominable

Aquinas's ideas about the passions formed the bedrock of the emotional regime under which most Europeans lived at the time of the witch crazes. To be sure, people challenged him, but they tended only to make subtle tweaks to his ideas, and the impact of their work rarely extended beyond their own limited readership. Written records from the era that discuss emotion, either directly or indirectly, tend to stay within the Thomist (as Aquinas's theories are known) framework, regardless of genre. His description of fear was what people would turn to to describe their fear.

There are even hints of Aquinas's passions in Shakespeare. In *The Rape of Lucrece*, for example, the Bard writes:

> *And extreme fear can neither fight nor fly,*
> *But coward-like with trembling terror die.*[9]

In the Thomist view, fear is a struggle that occurs when fleeing— in this case, the passion known as flight—is impossible. And this is important. Did you notice that both Shakespeare and Aquinas separate fear from fleeing? If you remember, modern psychology thinks that fleeing is an essential component of fear, alongside fighting and freezing. But to Aquinas and Shakespeare, once the opportunity to avoid harm has passed, the opportunity to flee has gone with it. You're in the thick of it, and that's when fear kicks in. Fear, back then, was not associated with fleeing. Fear, according to Aquinas and Shakespeare, is the opposite of bravery; it makes you coward-like. In some respects, Aquinas's fear sounds a bit like the emotion we refer to as panic today. This is why it's so important to make sure you know what historical figures mean when they use a seemingly straightforward word such as *fear*. Often, it's not as simple as it seems.

The other Thomist passion we need to explore is one that may have struck you as a little odd: flight or abomination. This, as the "or" implies, isn't one passion but two. Flight, as we've seen, is the desperate need to get away from something you'd normally associate with fear—a sort of running away so the thing you fear can't catch up with you. Abomination is what makes you want to run in the first place, and it's a bit more complicated than flight.

Abomination isn't entirely different from modern disgust. Like disgust, it's all about the yucky and horrid, the unpleasant to look at and the morally wrong. It's brought on by those things that might make you say "gross" or, if you were an English person living at the time, "fee," "fi," "fo," or "fum." (As an aside, if you ever read "Jack and the Beanstalk" to anyone in the future, say "Fee-fi-fo-fum" as if you're saying "Yuck, ew, ick, gross," and it'll suddenly make sense.) The term *abomination* could also refer to that which can contaminate, harm, or corrupt people it comes into contact with—generally speaking, if something was disgusting, it would likely be referred to as an abomination.

There are two crucial differences between abomination and modern disgust. First, abomination seems to share the same Latin root and was used to mean the same thing in most other European languages. This is because, as I also touched on when we visited the ancient Hebrews, "abomination" was the Latin Vulgate Bible's preferred translation of the Hebrew words used to describe the many revulsions God feels when people sin—the very feelings Saint Paul knew all too well. The Vulgate, as the primary Catholic Bible, was the bedrock of many people's faith in sixteenth- and seventeenth-century Europe and had been for nearly fifteen hundred years. Though most witch trials took place in Protestant towns and cities, the concept of abomination loomed large in Catholic and Protestant areas alike.

The second difference, the big one, comes from that biblical understanding of abomination. Having looked at thousands of documents from the period, I can tell you that the word *abomination* almost always occurs alongside, or at least pretty close to, words such as *God, Lord*, and *sin*. And this is not just true of religious texts; it's also the case with any sort of written work from the era. As it was to the ancient Hebrews, an abomination in the eyes of medieval Europeans was revolting in the eyes of God. But despite these differences between medieval abomination and modern disgust, there's an element that both have in common. Both the disgusting and the abominable are seen as contagious.

The type of contagion associated with both abomination and disgust is known as *sympathetic magic*. First suggested by anthropologist James Frazer in his 1890 book, *The Golden Bough*, there are two parts to this phenomenon. First, the law of similarity: "Like produces like." Second, the law of contact or contagion: "Things which have once been in contact with each other continue to act on each other at a distance after the physical contact has been severed." One experiment that demonstrated the effect of these beliefs was performed by psychologist Paul Rozin in 1993. He and his colleague Carol Nemeroff asked people if they would wear a sweater previously worn by Hitler. Almost everyone said no, and they continued to say no even if they were told that the sweater had been sterilized, if they were offered money, or if they were told that Mother Teresa had also worn the sweater. The fact that it had belonged to such an evil man gave people the sense that his evil essence was, somehow, contained in the sweater, that a mere concoction of thread and cotton could, in some way, infect anyone who put it on.[10] It seems that our brains aren't good when it comes to differentiating moral behavior from physical contaminants, so they treat both threats the same way.

Tests performed across the world have shown that preschool children believe that people and objects they care about—such as their favorite toys and teddy bears—contain a similar, though in these cases positive, "essence."[11] This doesn't change as we grow up. It's what propels many of us to believe in ghosts or support a football team over the course of decades—even though the only things the current band of traveling athletes shares with the players we first saw when we were kids is a team name and perhaps a logo or uniform. It's also why we pay through the nose for things just because a famous person has scrawled over them with a Sharpie.[12] What's important is that essences are always attached to an emotion. The associated feelings can be good, but often they aren't.[13] In the case of witches, sympathetic magic only reinforced their abominable nature.

How to Be a Witch

If you wanted to find a witch in the sixteenth and seventeenth centuries, you had look for someone who made you experience feelings of abomination. It would help if that person was a woman. Around 80 percent of the people executed as witches in Europe were women. (In some areas, including Russia, the opposite was true. Why that is we aren't entirely sure, but it may have something to do with the Orthodox Christian faith holding a different view of witchcraft.) Part of the reason for this was the deep misogyny that permeated the period—all of history, really—and, sadly, hasn't exactly disappeared today.

For a long period in Christianity, it was believed that all the evils in the world happened because Eve committed the biggest sin ever: disobeying a direct order from God. It was Eve who first ate the fruit of the Tree of the Knowledge of Good and Evil. It was Eve who persuaded Adam to do the same. It was Eve who got us all wearing clothes and cast out of paradise. It was Eve who was punished with the additional burden of menstruation—which was itself thought of as abominable—and childbirth. Some people thought Adam could have just said no to Eve and shouldered some of the blame. Ultimately, though, they argued that it was Eve who was won over by the serpent, and so the blame rested with her—and therefore all women.

Other people did a better job of defending women. They pointed out that the most exceptional human who ever lived was a woman— Mary, mother of God. But in general, women were thought of in ways that persist as the hallmarks of modern misogyny: weak, emotional, unable to control themselves. Nicolas Rémy, a French magistrate who wrote one of the most famous witch guides of the era, *Demonolatry*, claimed: "[Women are] more susceptible to evil counsels." He thought it was easier for the devil to turn women into witches than it was to turn men into witches because of women's tendency to be hot-blooded—literally, in this case, because an imbalance in the body caused by hot blood was believed to be the cause of bad behavior. So if you wanted to find a witch, it was easier if you started with women. Preferably passionate ones.

It also would help if those women were poor and old (by the standards of the time). Usually, people accused of being witches were "undesirables" living on the fringes of society who were over the age of forty. Popular targets were childless widows and strangers who had to beg for food and sustenance.

The feeling of revulsion supposedly came from a witch's "filth, stench, putrification," and "the menace they presented."[14] In reality, it more likely came from an all-too-familiar source: the objectification of women's bodies. The engraving on the next page, from around 1500, by Albrecht Dürer, *Witch Riding Backwards on a Goat*, is but one example among dozens of that objectification. Depictions of witches from the time almost all show the same thing—women who look old, with aged, imperfect, and infirm bodies; women whom male artists didn't find sexually attractive. In this way, what we now refer to as the male gaze was central to the process of making witches abominable.

Even so, some witches were young. To be a young witch required performing some sort of activity that God would find revolting. In another image from the time, Hans Baldung Grien's *Young Witch with Dragon*, we see a young, pretty witch performing abominable acts. This drawing depicts bestiality, sodomy, flatulence, the exchange of bodily fluids, "unnatural" penetration, and other sex acts between a witch and that most abominable of biblical creatures, the dragon of Revelation.[15]

Albrecht Dürer, *Witch Riding Backwards on a Goat*, engraving, ca. 1500

Witches were also often accused of attending a sabbat. Sabbats were an inversion of what was right and proper for Christians at the time. The *Compendium Maleficarum*, published in 1608, contains several images of a sabbat. These show witches stamping on the cross, performing mock baptisms, kissing Satan's buttocks, cooking and eating unbaptized children, and engaging in other "execrable abominations."[16]

Hans Baldung Grien, *Nackte junge Hexe und fischgestaltiger Drache*, drawing with body color (black-and-white reproduction), 1515

Woodcut from Francesco Maria Guazzo, *Compendium Maleficarum (Mineola, NY: Dover Publications, 1988)*

Looking abominable or performing abominable acts had the added effect of making people believe you could contaminate them with your evil. The sympathetic magic associated with abomination meant that witches could infect everything they touched, everyone they met, and every town they visited. They didn't have to actually touch anyone: looking sideways at people or cursing them out was enough.

The early modern mind believed that infections could be transmitted via insulting words or by sight alone. Witches were regularly accused of using a look, or the evil eye, to harm not only the people they despised but also the judges at their trials.[17] This might seem odd now, but the firmly held beliefs of the time made it a powerful tool, potentially inflicting real psychological damage. In addition to the power of sympathetic magic, there's a phenomenon in psychology called the nocebo effect. It's the opposite of the better-known placebo effect. When people are sure something is going to harm them, it appears (at least in some studies) that their physical or mental health suffers.[18] So if you absolutely believe in the power of the evil eye, it actually can hurt you—though in a sense, you're really just hurting yourself.

In those fear-dominated times that were the sixteenth and seventeenth centuries, it didn't take much for a woman, especially a poor and old one, to be accused of being an abominable witch. You needed only to fail to conform to the prevailing definition of beauty, be old, be less reserved than was considered proper, and look at people the wrong way. The message to women who wanted to avoid the horrors associated with a witchcraft accusation was clear: stay young, stay pretty, don't be too emotional (unless you're filled with the emotions of fear, abomination, and the urge to kill witches), and do as you're told.

Maybe one day we'll stop saying that.

How to Stop Our Witch Crazes

The impact of the witch crazes on history is complicated. For a start, they highlighted an era of ever-growing fear and deep misogyny. But they also shine a light on the modern world, because witch trials are not just a thing of the past. Later in this book I'll discuss modern, metaphorical witch hunts, in which people on all points of the political spectrum are attacked, dogpiled, and driven out of jobs and off social media. Usually, this is the result of yet another aspect of disgust—its relationship to ideological purity. But sadly, not all modern witch hunts are metaphorical.

The prosecution, torture, and execution of people accused of being witches still goes on in India, Papua New Guinea, Amazonia, and much of sub-Saharan Africa. In Tanzania alone, an estimated forty thousand people were accused of witchcraft and killed between 1960 and 2000. In the United Kingdom, a fifteen-year-old boy named Kristy Bamu was tortured and murdered during an attempted exorcism by his sister and her boyfriend in 2010. And the same emotions that ran rampant during witch crazes in the distant past are at play in these modern cases: the idea of sympathetic magic, in which the accused is able to harm not only by touch but also by a look, and the abomination of the idolater, whose ungodly ways bring calamity onto the people who oppose them. What makes it somehow worse is that in many parts of the world, the accused are children, sometimes young children.

In a period of history filled with horrors, the witch crazes—and the uncontrollable fear that bolstered them—turned out to be one of the worst terrors of all. As for how that shapes the modern world, perhaps remembering and understanding the forces that were at work back then can help us mitigate some of those same forces today. Now, as then, most people who accuse others of being witches—metaphorically or otherwise—aren't necessarily evil. They are afraid and so lost in superstition that they would do anything to reduce that fear. We need to assimilate the lessons of our emotional past. Find ways to educate people who, because of their dread, would do others—including children—harm. The European witch crazes offer a lesson we ought to learn. Remembering them, and studying them, can teach us much about the witch hunts, both metaphorical and real, that are happening right now.

Seven

A Desire for Sweet Freedom

Samuel Adams had been trying to keep control of a particularly riotous meeting for quite some time. It was a cold December day in Boston, Massachusetts, but inside the Old South Meeting House, tempers were running high. Adams understood everyone's frustrations. As the local Whig Party leader, he had opposed the taxes levied on luxury goods ever since they were introduced. But the British Parliament seemed to think that heavily taxing objects of desire was a good idea. Many of the people living in America agreed with Adams, but their objections were fruitless. They had no say in the matter, and that infuriated them.

Very few people like paying exorbitant taxes on products they enjoy, but to understand why colonial Americans were so furious about the taxes imposed on them by Britain, we need to know a bit about their conception of justice. America's founders were deeply influenced by Enlightenment thinkers, and they were particularly enamored of the concept of natural rights, first posited in the work of the seventeenth-century philosophers Thomas Hobbes and John Locke. The basic principle is simple enough: natural rights are rights to which everyone is inherently entitled. They don't come from God, and they don't come from kings. Locke fashioned them into three "inalienable" rights: life, liberty, and property, or the ownership of all you create.[1] In fact, Locke's formulation was so deeply influential in the founding of America that it appears in the first line of the Declaration of Independence, albeit with the word *happiness* substituted for *property*.

The leaders in Boston and the rest of America were educated men. They'd read Hume and others who argued, among other things, that "by rendering justice totally useless, you thereby totally destroy its essence, and suspend its obligation upon mankind."[2] They had read John Locke and knew that their natural rights were being violated by the British government. Here stood a well-educated people, their sentiments burning at the continued immoral oppression being meted out upon them.

But it was the Tea Act of 1773 that seemed to push them over the edge. This bit of legislation introduced a sly tax hidden under a new deal on tea imports and handed a monopoly on those imports to one organization: the British East India Company. It was an attempt by the British government to get rid of an excess of tea owned by the company, which was suffering terrible financial difficulties at the time. The idea was that by edging out the tea coming in through the black market, the company could sell its excess stock and make some money. The problem was, this meant that the people buying the tea had to pay yet another tax to the British government. Without any representation in Parliament, the citizens considered this a bridge too far. In challenging this tax, colonial Americans saw an opportunity to show Britain exactly how angry they were about its ever-increasing infringements on their natural rights.

Following the passage of the Tea Act, most colonies began turning British East India Company tea imports away, and some forced the officials who accepted tea imports to resign their positions. But the residents of Boston found themselves in a particularly difficult situation. The sons of their governor, Thomas Hutchinson, were in the tea trade, so he staunchly refused to participate in the boycott. When a ship called *Dartmouth* arrived in Boston Harbor, a resolution was passed to force it to sail back to England. Hutchinson ignored the decision. He knew that if he stalled for twenty days, he could legally unload the cargo onto American soil without the approval of the local council. Then two more ships carrying tea landed at Boston.

The people of the city were outraged. And this was why Adams called for an assembly at the Old South Meeting House. His call was heard—and then some. Some reports claim that as many as half of Boston's sixteen thousand residents crammed into the hall.[3]

What exactly happened next is debated. Some hold that it was luck, others that it was a brilliant plan on Adams's part. But regardless of his intentions at the time, when he announced that "this meeting can do nothing further to save the country," most of the attendees left, and they left furious. Depending on whom you ask, somewhere between 30 and 130 of them went home, dressed up as members of the Kanien'kehá:ka tribe of Native Americans (also known as the Mohawk), and made their way to the seaport. They proceeded to board the ships and dump 342 chests of tea into the ocean. In a matter of hours, a great deal of luxury, time, and money became food for the fish.

But Britain wasn't going to allow its colonists to cut the cash flow quite so easily. King George's government retaliated by passing a series of laws that have become known as the Intolerable Acts. They stripped Massachusetts of its right to self-governance, moving it further away from having the representation the citizens wanted in Parliament. They closed the Boston docks until the tea had been paid for. They made it challenging to put royal officials on trial in Massachusetts, regardless of the crime, by increasing the cost of pressing charges beyond the amount most people could afford to pay. They also made it legal for British soldiers to be housed anywhere the state governors saw fit, including in private homes, though it's unlikely this ever happened. The colonists saw these acts as an unforgivable violation of their natural rights, an act of direct oppression by the British Crown. The American War of Independence had become an inevitability.

*　　*　　*

That's a lot of fuss over a few cups of tea, you might think. But it wasn't really about the tea. It was about the taxes and what they represented. It's a beautiful example of the power of raw anger. What caused that rage—taxes on luxuries—shouldn't be overlooked. I've already mentioned the power that the desire for luxury can bring when I related the story about the Ottomans blocking the Silk Road, forcing Europe to find other routes to the East. Without the desire for nice things, who knows when the continent on which the Boston Tea Party unfolded would have been discovered? But rather than exploring the anger of the oppressed, something that, while important, is a little bit obvious, let's take a look at the broader implications of a desire for luxury, a desire so potent that to overtax it can lead to rebellion, the rise of a new nation, democracy, rock 'n' roll, deep-dish pizza, moon landings, and Hollywood movies (to name but a few things).

We know that the fall of Constantinople to the Ottomans led to a search for new trading routes. But despite the Ottoman tariffs, the riches discovered in the New World and the route to the East used by the Portuguese and the Dutch East India Company increased the influx of luxuries as never before. The desire for *things* that this uptick in trade unleashed would profoundly shape history and pave the way for the modern Western world.

Becoming Tasteful

The desire for material things, or *greed*, existed long before the Ottomans invaded Constantinople, of course. It could be easily argued that many crusading Christians, Alexander the Great, and even Ashoka did what they did, at least in part, out of greed for power, fame, and fortune. When conquering worlds, it's hard to separate wealth from power. But back in the times of Ashoka, the Crusades, and Alexander, people were warned continuously against indulging "first-order desires."

If you recall our earlier exploration of desire, these are the selfish desires for riches, for possessions, for personal wealth. You might call first-order desires *greed*.[4] It's better to embrace second-order desires: the desire to desire, or, perhaps, as is the case in Buddhism, the desire not to desire. Second-order desires help you control your feelings and achieve something better than material gain: a virtuous life, *eudaimonia*, *nirvana*, and heavenly rewards.

But wealthy Europeans in the eighteenth century wanted to buy nice shiny things *and* go to heaven. It didn't take long for philosophers to find a way to square that circle, and they did so using a different category of feelings—*sentiments*, specifically, the sentiment of taste. That's right: taste was thought of as a type of feeling. To explain why, we need to delve into the long history of taste and sentiments.

Versions of the word *taste* began appearing with increasing regularity in Europe in the early sixteenth century, right around the time that the number of luxury items entering the continent started to rise. Obviously, people knew what taste was, and they had for quite some time. If we go back to our old friend Aristotle, we discover that taste was considered one of the lower senses. It, along with touch and smell, meant touching something in nature with a bit of the body. Sight and hearing were better, higher senses. Keep in mind that Aristotle, for all his brilliance, hadn't the first clue about electromagnetic radiation or sound waves. We now know that the perception of color occurs when our brains decipher the light reflected off an object after it enters our eyes. But Aristotle thought that color was part of an object—something *in* it. He took this to mean that sounds and sights went straight into the soul without touching anything. After all, most people don't have to stick something in their eyes to see it or in their ears to hear it.[5] But taste means touching an object with your mouth, with an orifice, and we humans tend to get a bit squeamish when it comes to our holes.[6]

Taste could do one thing the other senses couldn't: tell the truth, at least to some degree. What tastes bad tastes bad, and that's that. By the tenth century, taste was still considered a low, crude sense. The tenth-century Benedictine abbot Saint Anselm of Canterbury included enjoying flavors in his twenty-eight sins of curiosity. Taste also played a role in more widely denounced sins such as gluttony and sex, because both, when done well, involve the mouth.[7] Before long, debates about the virtue and vices of taste began. Entering this dispute was a new group of thinkers—the humanists.

This type of humanism (Renaissance humanism) is not to be mistaken for secular humanism. Unlike secular humanists, Renaissance humanists were very much believers in God and the Church. Their movement began in Italy in the fourteenth century and played no small part in energizing the period in history from which their name is derived: the Renaissance. As a group, they are hard to define. But if anything united these rather disparate individuals, it was a pursuit of lost knowledge and a desire to return to a classical golden age. They wanted to go back to a past found in the ancient works of philosophy that began surging into Europe from the East during the twelfth century. To do this, they read and translated books they were convinced were from that golden age—from back when humanity was closer to creation and so less corrupted by sin.

Through those original texts, they rediscovered the Roman focus on this life rather than the hereafter. The movement had an effect on art, inspiring images known as memento mori. These were reminders of mortality—skulls, images of death, that sort of thing. They were meant to remind the viewer that death comes to us all, regardless of our wealth or power.

Another aspect of the humanist attempt to find a more perfect, less corrupted past and bring it to the present involved an exploration of ancient beliefs about morality, food, and taste.[8] In ancient Rome, activities related to food production, such as farming, selling, salt-curing, drying, and a range of other techniques (sausage was invented by the Romans, for example) were considered dishonorable, close to

Hans Holbein the Younger, *The Rich Man; The Queen* (1523–43), woodcut *(National Gallery of Art, Washington, DC)*

flesh, dirt, and the earth. It was not something for a refined Roman citizen to get involved with. Many of the Roman works the humanists read described immoral acts of lust and desire alongside mentions of food producers such as fishmongers and butchers.[9] The humanists adopted this outlook. But humanists, like Romans, had no issue with food once it had been cooked or with those who cooked it, especially if it had been cooked well. Some humanists, including Bartolomeo Sacchi, or Platina, as he's better known, went so far as to link well-prepared food with a healthy moral life.

Sometime around 1465, Platina bought a copy of *Libro de arte coquinaria* (*The Art of Cooking*), written by the great-great-godfather of every modern TV chef, Maestro Martino.[10] Though not the first cookbook by any stretch of the imagination, it was the first taken up by other chefs en masse. Even housewives with a little spare money would splurge on a copy. Platina used Martino's recipes in a book he published around five years later: *De Honesta Voluptate et Valetudine* (*On*

Right Pleasure and Good Health). Platina provides an excellent example of the way the humanists linked taste—or, rather, good taste—with health and virtue.

Only a few years later, taste was used as a metaphor for getting close to God. In England, a translation of Gui de Roye's ca. 1489 book, *Le doctrinal de sapience*, warned:

> There be many Christians, both clerks and laymen, who little know God by faith nor by scripture because they have had their tastes disordered by sin, and so may not savour him.[11]

God only tastes terrible if you are a sinner, apparently. Sadly, there are no records that reveal how food tasted to the wretched. But we might assume, from Platina's work, that what they ate made the dishonest unhealthy.

To the humanists, taste was a way to know something, but it went further than just flavor in the mouth. Taste was something you could use to evaluate the good and bad in art, poetry, prose, even behavior. Cultivating one's taste for the beautiful was an essential part of living a good life. Taste became a metaphor for a new right sort of desire. Well-selected luxury, and the ability to recognize what real beauty is, was suddenly considered a force for good. The humanists didn't go as far as to make the ownership and appreciation of certain luxuries an absolute moral good, but they certainly laid the groundwork for that belief. To see how that shift happened, we have to jump forward a few centuries—to the Enlightenment.

Accounting for Taste

Back when the humanists were thinking about what constitutes good taste, luxuries were, on the whole, expensive and rare. Only the elites had pepper, silk, and china plates. By the early seventeenth century, however, prices had dropped and wages had risen. Education across the board had improved. Literacy had increased, and books were inexpensive enough to be within reach of almost everyone.[12] The nobility, who enjoyed feeling that their baubles and shiny things were rare and exclusive, were upset. They didn't like the fact that prices had dropped so low that European knockoffs of their luxury goods could be sold to people they viewed as beneath them.

The upper classes, then as now, liked to distinguish themselves from the poor. One early seventeenth-century philosopher, Bernard Mandeville, thought this was a good thing because it drove commerce and innovation.

His argument went like this: a rich woman might buy a lovely dress in the latest style—perhaps made of imported silk from China and gold braid from Africa. Someone slightly less well off would see the dress, admire it, and buy a knockoff version—maybe satin with yellow braid—from a European dressmaker. Soon the popularity of this design would lead to ever-cheaper versions until all but the poorest women could be seen out and about in something frighteningly similar to the wealthy lady's gown. Seeing poor women wearing her expensive dress would cause her to suffer from something Mandeville called odious pride. Odious pride was good on the whole because it would force her to seek out new clothing, encouraging designers to innovate and create new styles. On and on the cycle would go.[13] It's a cycle that still happens today. How often is an exclusive outfit, worth thousands of dollars, worn by a celebrity only to be copied and sold in a chain store just a few weeks later? These days, we call that cycle *fashion*.

Of course, the Church was not happy. Clergymen weren't humanists, for the most part, and they hadn't come around to the idea that taste could be a good thing. Wanting pretty dresses, shiny baubles, and beautiful tableware seemed to be distracting people from what mattered most—their path to God. To the dismay of the clergy, the desire for earthly objects seemed to be outstripping the desire for an afterlife. That didn't mean that people had forgotten about God and the hereafter. Far from it. But, like the stereotypical Ferrari-driving pastor of an American megachurch, they sought ways to reconcile their faith with owning beautiful things. Thankfully, there were plenty of philosophers who, like Mandeville, were happy to find a way to make that pious square fit into a nicely decorated round hole.

In the early eighteenth century, a group of English philosophers took up the cause of good taste. They developed it further than anyone had previously, including the Renaissance humanists. One of these pioneers was Anthony Ashley Cooper, 3rd Earl of Shaftesbury. His views on taste became a central point of debate for everyone who wrote about taste and aesthetics for the following one hundred years. In his 1711 book, *Characteristicks of Men, Manners, Opinions, Times*, Shaftesbury, as most historians call him, said that beauty comes from objects with certain qualities—harmony, order, symmetry, proportion, design, and numbers (as in not having so many of something that it renders an object gaudy).[14]

The chair of moral philosophy at the University of Glasgow from 1729 until his death, in 1746, Francis Hutcheson owed more than a small intellectual debt to Shaftesbury. Although the two men were contemporaries, they never met. But when it came to ideas, they were almost inseparable. Hutcheson's work combined ideas of morality and taste in almost exactly the same way Shaftesbury did, so much so that he included Shaftesbury as coauthor on some of his early works, even though they weren't collaborations. To Hutcheson, good taste was about the just-right balance of component parts. The issue was that one person's "just right" was different from another's. The reason for this was that some people simply weren't refined enough. He wrote:

> *Bad* Musick pleases *Rusticks* who never heard any better . . . A *rude Heap* of Stones is no way offensive to one who shall be displeas'd with *Irregularity* in *Architecture*, where *beauty* was expected.[15]

The poor and uneducated, according to Hutcheson, only like noisy folk music and ugly houses because they don't know any better. If only there were a way to cultivate good taste. Well, he thought there was: by improving your education, you could learn to refine your taste so that you would react appropriately to that which is genuinely beautiful. It was all about learning how a given objet d'art is supposed to make you feel.

Oh, and just as a quick aside, it's this group of thinkers who started to use the word *disgust* as we now understand it. To them, disgust was a sort of anti-taste—a reaction to something ugly, unpleasant, and aesthetically wrong. But let's get back to taste and how it fits into the concept of *sentiments*.

Becoming Sentimental

The Enlightenment is often portrayed as an age of reason, a time when thoughts were much more important than feelings. Feelings got in the way of objective truth, which is the thing Enlightenment philosophers were most preoccupied with. Some even went so far as to invent mathematical equations to explain topics as diverse as the existence of God, law, politics, and ethics.[16] But that's only part of the story. A better way to understand the Enlightenment is as a period of secularization,[17] an era when the slow realization that religion might not have all the answers began to dawn on even people not previously brave enough to say so. That shift gave rise to new ways of thinking about morality, including the category of feelings that Enlightenment thinkers called sentiments.

Sentiments were described by Shaftesbury as a "moral sense." Think of the anger, disgust, and outrage many felt during the storming of the US Capitol by insurgents on January 6, 2021. Alternatively, try to imagine, or remember, the pleasant sensations of joy you get when seeing someone help an old lady across the street. Today, in accordance with modern science, we call these sorts of feelings emotions. But Enlightenment philosophers referred to them as sentiments—the sorts of feelings you have when someone does something you consider to be good or evil.

There was disagreement about the way sentiments worked. Before the Enlightenment, philosophers would probably have viewed them through the prism of God. We've covered quite a few understandings of emotion that tie how we feel to notions of sin. But that explanation wouldn't do during the Enlightenment. Instead, Shaftesbury thought they were simply part of our intrinsic physical nature, much like taste. To Hutcheson, they were a reaction to the characteristics of balance and imbalance. A young person helping an old lady across the road shows a sort of balance. Young and old, frail and healthy. Similarly, people storming the US Capitol would be viewed as upsetting the delicate harmony of American democracy. David Hume put it down to utility. If I help an old lady across the road, that allows the old lady to get to her destination safely. I feel good about helping her, because helping her is *useful*. If I ignore her and she is hurt on her way across the street, that is not useful, so I feel bad. Another famous sentimentalist, Adam Smith, explained sentiments in terms of sympathy. He thought we all have an "impartial spectator" living within us who feels bad when other people feel bad and good when other people feel good.[18] We might now call that empathy.

Sentiments came in two flavors, or at least that's how I and most modern historians tend to look at them. The first flavor was moral sentiments, as described above. The other was the one that links to taste—what we might call aesthetic sentiments. These are the feelings that tell us what is beautiful and tasteful or disgusting and tasteless. On the face of it, these two categories of sentiments don't seem to have anything to do with each other. After all, knowing that a BMW is pretty doesn't make you a good person. Or at least, it doesn't now. But people in the eighteenth century didn't draw this binary. They knew only of *sentiments.* Adam Smith, for example, wrote about "beauty of every kind," meaning not just aesthetically pleasing beauty but also "the beauty of conduct" as well.[19] The sentiments for the morally wrong and the aesthetically unpleasant were identical in the minds of the Enlightenment thinkers; they produced the same feelings. To be tasteful suggested that you were moral and sound. To be tasteless suggested otherwise. Beauty was good; evil was ugly. That's how the square was rounded and the desire for things became virtuous. It was okay to covet as long as you felt the right sentiments. In other words, material desires were fine, so long as you had good taste.

Cash and Cannons

The fact that the sentiment of taste allowed desire to be seen as a moral good shaped history in profound ways. First, it helped form the modern state as we know it. Second, it was crucial to the development of modern capitalism. For evidence of my first claim, let's take a trip to the Netherlands.

One day, if you are lucky enough to do so, take a leisurely walk through Amsterdam. On the road that follows the Keizersgracht, or Emperor's Canal, you'll find the Huis met de Hoofden, or House with the Heads. A beautiful building that houses a museum, it takes its name, as you might have guessed, from the ornamental heads that adorn its exterior: Apollo, Ceres, Mercury, Minerva, Bacchus, and Diana. This home was once the residence of Louis de Geer, who purchased it in 1634. On the banks of another of the city's canals—the Kloveniersburgwal—sits another beautiful building. The Trippenhuis is a mansion built in 1660 to reflect the classic tastes of its owner. It's festooned with decorative weapons of war. Cannons and cannonballs are carved into its stone facade alongside a symbol for peace—an olive branch. These are two stunning homes, and both are temples to the wealth of their entrepreneurial owners. These were the homes of arms dealers who built and supplied weapons for the Dutch government and the East India Company. Their services were thought to be essential. Without weapons of war, there could be no peace. Without peace, the stream of luxuries might cease. But going to war over so grand a prize was inevitable. Competition for the riches of the Americas and the Spice Islands was fierce. England and the Dutch Republic came to blows on no fewer than four occasions over the wealth that colonial luxuries provided: once from 1652 to 1654, once from 1665 to 1667, again from 1672 to 1674, and finally a century later, between 1780 and 1784. The Dutch won the first three but lost the last to the newfound naval might of Britain.

Protecting commerce and, more important, the taxes that come from trade was a good reason for war. The problem was that, initially, most countries had no standing armies or navies. They still relied on recruiting soldiers via the nobility as needed. Something more organized was necessary—a paid and trained fighting force, ready to do battle on land and sea at all times. Funding for such ventures came from each country's taxes, and so a feedback loop was created. Armies and navies were needed to protect the traders who paid the taxes that funded (among other things) the armies and navies. It was a relationship between tax and fighting that has come to be known as the fiscal-military state. Most of us now live in some form of that state, and it's all thanks to the way we adjusted our moral and philosophical outlook to accommodate that voracious form of desire: greed.

As for my second claim, we've already explored the role desire and taste played in the creation of fashion. Not much has changed. Desire and taste continue to sustain contemporary consumer culture to this day. Desire drove competition, built the European knockoff shops, and kick-started social mobility. The history of capitalism is long and very complicated, so I won't go into that here. Suffice it to say that none of it could have happened if people hadn't coveted their neighbors' goods, desired trinkets, and managed to come up with a way to *feel good* about doing so. First-order desires and the sentiment of taste unlocked commercialism and made free-market capitalism into a moral good that might lead to happiness. To paraphrase American debater and writer Matt Dillahunty, the United States was built by breaking the Tenth Commandment: each person coveting his or her neighbors' goods.

But this hasn't always been the case, at least not in such simple terms. The desire for liberty and self-governance had overridden the desire for luxuries in Boston when the citizens dumped the tea. A similar drive soon engulfed many other colonies in North America. The British government had miscalculated. The Americans were not going to pay a tax they had no say in, on tea or anything else. They'd rather do without, as demonstrated in no uncertain terms by the Boston Tea Party. Freedom was the ultimate luxury. It was beauty, truth, and a moral good.

As I mentioned above, America had three inalienable rights of its own enshrined in the Declaration of Independence. Like Locke's, they guaranteed everyone the right to life and liberty, but the founding fathers changed Locke's right to property to the pursuit of a passion: happiness. In their eyes, happiness and property had plenty in common. The pursuit of joy had underpinned systems of feeling outlined by the likes of Plato and Aquinas. Now the right to happiness was expressed in the foundational document of a new nation. But this time, and for the first time, it was intimately tied to desire. The sentiment of taste and the desire for freedom, wealth, and luxury had birthed the United States. It was to build itself into a nation revolving around the pursuit of happiness — happiness fueled by the desire for riches.

Eight

Becoming Emotional

So far, we've been examining types of feelings that aren't quite the same as the feelings we, as modern people, recognize. We've considered *pathē:* perturbations in the soul brought about when the senses detect something that might cause pleasure or pain. We've considered *passions:* feelings very similar to *pathē* that are felt in the sensitive part of the soul and that can influence the rational soul if you let them. And we've considered *sentiments:* feelings that help us with moral actions or aesthetic judgments. But none of these quite lines up with the modern idea of *emotion.* While the previous few chapters have discussed specific emotions within the framework of these larger categories of feeling, this chapter will look at the bigger picture. It will explain how emotion — the category of feelings modern psychology all but takes for granted — became a *thing.* The story of the birth of modern emotion involves a bit of philosophy and a bit of science. In fact, the person who kicked off the change from older ideas of feelings to the modern idea of *emotion* just happens to be one of the most famous philosophers of them all.

I Feel, Therefore I Am

In 1650 — just over a century before the concept of emotion came to mean what it means to modern psychology — a fifty-three-year-old

Frenchman lay dying of pneumonia in a cold, damp house on Väster-långgatan, one of the main streets of Stockholm, Sweden. Although his curly black wig had been removed to reveal thin gray hair, the sharp mustache and pointed goatee that would make him instantly recognizable for centuries to come were still a pristine black. Just a few months earlier, René Descartes had taken the job of tutor to Queen Christina of Sweden, moving there from the Netherlands, where he had lived most of his life.

Queen Christina had really liked Descartes's published letters to the rather brilliant Princess Elisabeth of Bohemia. She loved them so much, in fact, that she persuaded him to turn them into a book, *The Passions of the Soul*. Christina especially seemed to have liked the fact that Descartes approached feelings, as he put it, not "like an Orator, or Moral Philosopher, but as a physician."[1] Despite dipping their toes into Galen's ideas about emotions and the humors, books about the passions in the seventeenth century tended to be based on Aristotle's *Rhetoric:* like books about oratory, they were typically about ways to use feelings to change people's minds. Those that weren't updates of *Rhetoric* were books about manners: many moral philosophers placed control of the passions at the center of their guides to good behavior in the late sixteenth and early seventeenth centuries. Decorum, especially in business and diplomacy, had become a central element of the emotional regime that governed European society. The sword had been replaced by bowing—at least in theory. Descartes's dedication to exploring emotion somewhat scientifically was a novel approach.

Despite Christina's fondness for Descartes's work, becoming her tutor ultimately didn't work out so well for him. It had sounded like a great job at the time. Europe had finally emerged from eighty years of near-constant war, making traveling much safer and easier. The job paid good money and provided relative safety from the people who didn't like him. It also lent him the ear of royalty. What Descartes almost surely didn't realize was that he wouldn't be teaching Christina about feelings but instead would be teaching boring old ancient Greek. He had no clue that he wouldn't get along with her very well. And he definitely did not expect to be giving the lessons in a freezing old castle that would ultimately kill him.

As Descartes lay on his deathbed, he would have been able to look back on his life with pride. As a young man, he had been a decorated military engineer. He had transformed geometry by showing the world how to measure lines using coordinates and equations. He had developed an accurate model of the colors of the rainbow. He had also nearly beaten Isaac Newton to the punch with an almost-but-not-quite-right version of the laws of motion. After those significant achievements, he went on to upset nearly two millennia of academic thought by inventing an entirely new type of intellectual inquiry called foundationalism. He had revolutionized philosophy and science forever. He was something of an overachiever, it's fair to say.

A great deal of ink has been spilled on the subject of Descartes's work and the influence it had on the thinkers who followed. Summarizing all his accomplishments here would easily fill the remaining pages of this book and then some. Thankfully, we don't need to work our way through all that. But to understand his views of emotion, we do, at least, need to explore two of his most important beliefs.

The first concerns his views on the nature of the material world. In 1618, while serving in the Dutch States Army, Descartes met Isaac Beeckman, a natural philosopher and proponent of an increasingly popular idea: that the universe and everything in it behaved mechanically, like clockwork. Beeckman accepted that the earth was not the center of the universe and that everything in this sun-centered, mechanical universe was made up of tiny atoms, or corpuscles. It was these atoms bouncing off each other that made the universe's clockwork tick. But these weren't quite atoms as we understand them today. Beeckman didn't know about electrons and protons. He had no clue about nuclear forces and electromagnetism. Atoms were just tiny bits of variously shaped stuff that made up bigger bits of stuff. Descartes took these new ideas and ran with them. They formed the basis of much of his late work, but for our purposes, the most important thing to note is that he bought into the idea that the physical universe is fundamentally mechanical.

The second of Descartes's beliefs that's relevant to the birth of emotion appears in his most famous book, *Meditations on First Philosophy*. In fact, it is summarized by its best-known line: "I think, therefore I am" (or "Je pense, donc je suis" in the original French). Descartes coined this simple phrase as part of his effort to prove a bunch of philosophers wrong. Not long before, a group we now call the Christian skeptics had concluded that you can't really know anything. They believed this because, unlike God, we have limited brains. You have to take the existence of God and everything else on faith. But Descartes wanted to know if there was anything you could actually know for certain—a properly basic belief, or a foundation on top of which you could build everything else. "Je pense, donc je suis" was his answer. He thought that because he could think the phrase "I can think," then something must exist that thinks the thought "I can think." It didn't matter if that thinking thing was a brain in a jar, a creature having a dream, or a real human being. It was still a thinking thing, and it was still "him." God, he mused, wasn't needed in order to know something that foundational.

That's not to say Descartes was an atheist. Far from it: he used his thinking-thing foundation to build up a couple of arguments for the existence of God. He just thought that because God made the universe so it would run like clockwork, he didn't need to interfere with anything after creation. Descartes also thought that you didn't need faith alone to believe in God. You could use foundationalism to step from proof that you exist to, after a few strides, proof that God exists. But to the Catholic Church, these ideas made him as good as an atheist.

Fearing a backlash, he ran off to the Netherlands, a largely Protestant country at the time.

Passions of the Soul

Putting to one side the matter of whether Descartes's arguments for the existence of God hold up under scrutiny, in the process of forming them, he did something of vital importance with regard to our ideas about emotion: he got rid of the Greek three-part soul. He came to believe that humans must be made of two parts—the material body and the immaterial soul. The first was the mechanical and natural bit that dies; the second was the thinking thing. A staunch believer in the afterlife, Descartes used this formulation to explain how the thinking thing persisted after the mechanical bit died. He even pinpointed—incorrectly—where these two parts interact: a tiny part of the brain called the pineal gland, which sits at the top of the spine by a lump known as the thalamus, an area that we now know helps coordinate things such as our senses. Descartes thought that when atoms inside the body are moved by atoms hitting the sensory organs—such as the eyes, the skin, the nostrils, and the inner ear—they create heat, hunger, thirst, dryness, wetness, and a whole host of other "nesses," which in turn move the pineal gland. The soul or mind then interprets these movements as the sensations of touch, sight, smell, and so on.

"Hang on," I hear you cry. "What about that book on feelings—*Passions of the Soul* or something—you mentioned earlier? Shouldn't that be what you talk about here?" Well, yes. But knowing *how* Descartes did his thinking is essential for three reasons. First, his mechanical ideas, borrowed from Beeckman, convinced Descartes that feelings were part of a mechanical body. Second, the way he understood feelings is tied directly to his method of knowing things through foundational beliefs. Third, in his book about feelings he used a relatively new word to describe a significant number of them: *emotion*.

For Descartes, the foundational passion—the passion from which all other passions derive—was admiration. As I'm sure you've come to expect by now, Descartes's admiration didn't mean admiration as we understand it in the present. The 1649 French version of admiration meant something like acknowledging that something exists and deciding how you feel about it. This makes sense. To feel something about something, there has to be something to feel feelings about, something to admire. After admiration come the feelings we use to make judgments: love, hate, desire, joy, and sadness. So when Descartes picked up a ball of wax that had fallen off a candle and thought about— or "admired"—it (which is something he wrote about doing in *Meditations*), he might love it because it looks nice or hate it because it burned him. He might also desire to possess it because it could keep his fabulous mustache and goatee smooth and pointed, or he might reject it because it would turn his signature facial hair the color of the candle. This, in turn, could make him either happy—what, I ask, makes one more joyful than a beautiful mustache and goatee? —or sad.

This is the significant bit in our story: Descartes believed that all passions were coupled with something he called emotion. The word had been around for approximately one hundred years. In both French and English, it meant something like a "commotion" or "turbulence" in the body. But Descartes changed the meaning in a tiny but significant way. To him, the turbulent stirring was caused by atoms hitting the sensory organs, which in turn moved more atoms inside the body to the heart, the blood, and various bodily fluids. Our senses admire something, and that admiration causes the heart to begin "e-motion," a motion outward. From there, atoms start rushing around, causing us to feel things by making our bodies react in specific ways—blushing, laughing, trembling, and so on. He said that "[un]til this emotion cease, [the passions] remain present in our thoughts."[2] It is this act of e-motion that hits the pineal gland, disturbing our souls like a pebble hitting a still pond. The disturbance makes it seem as if the passions are felt by the soul when really they're just movements caused by e-motions in the body. This isn't the definition of emotion that psychologists use today, obviously. But it is *an* understanding of something called emotion. To get from the Cartesian version of emotion to the one psychologists argue about in modern times, we need to look at one of Descartes's contemporaries, the philosopher Thomas Hobbes.

Taming the Leviathan

In 1651—the year after Descartes's *Passions of the Soul* hit the presses— Thomas Hobbes published a hugely popular book called *Leviathan*. In it, he suggested that passions arise when atoms hit the sensory organs, causing motion inside the body. Consider sight, for example. An image hits the eye, then the eye sends corpuscles—those same corpuscles Descartes believed in—to the brain, then the brain moves to process the information. For Hobbes, as it was for Descartes, it was all about motion. It was all mechanical.

Hobbes thought these motions—he called them *endeavors*, a word he used to mean a sort of straining or forcing caused by other motions—start off so subtly that you don't know you're experiencing them. Eventually they build up until you do notice them, and at that point they produce one of two raw feelings: appetite (or desire) and its opposite, aversion.[3] These are not unlike feelings we've come across before—the ones that are often placed at the center of a category called the desiring passions, including desire itself and flight or abomination. Hobbes knew this, of course. He would have read Plato, Aristotle, and Aquinas. But he chose to make desire and aversion into primary passions. Hobbes was also aware how important two other desiring passions, pleasure and pain, were to Aquinas and Aristotle, but as we'll see, he had a different job for those feelings.

At its simplest, in Hobbes's view, appetite moves a person or animal toward something. Aversion is the opposite. Any animal can feel appetites and aversions, but the more sophisticated a human gets—the better educated and more experienced—the more complicated his or her passions get. For example, you might feel *"Aversion* with an opinion of *Hurt* from the object," or, rather, the need to avoid something that can cause pain.[4] Hobbes called that sort of aversion fear.[5] But humans and animals alike can learn that sort of fear. For complicated fears, such as the fear of the unknown, you need a human mind. According to Hobbes, these more complicated fears came in two flavors: religion when the fear was condoned by society and superstition when it wasn't. If the object of a socially condoned fear was something genuine, he called that fear "true religion."[6] This got him into a lot of trouble: quite a few people labeled him an atheist because of this view and because at no point in the book does Hobbes say that God is actually needed.

Hobbes also said something more groundbreaking concerning emotions than is often acknowledged, and that is where his use of the words *pleasure* and *pain* come into play. Up until this point, I've said that feelings were judged as good or evil based on the ways they were used. The consensus in almost every discussion of emotion—from Plato to Buddhism to Aquinas—was that things like sinning and excessive amounts of sex, wealth, and material possessions (good taste excepted) are bad, even though they make you feel good. On the flip side, doing as your God dictates or what is virtuous, such as fearing God in Islam and following *dharma* in Hinduism, is good, even if it makes you feel bad. But Hobbes didn't separate the passions like that. He categorized them in the same way that most modern psychology does—by how they make you *feel*. If they make you *feel* pleasure, he called them good; if they make you *feel* pain, he called them evil. It didn't matter if you felt pleasure while sinning or pain while praying.[7]

The critical part of all this is that Hobbes wasn't writing a book about the passions. The passions simply played a role in his broader agenda, and that was to prove that people need a king—or at least something like one. As an Englishman, he'd lived through a lot—the death of Queen Elizabeth I, the Gunpowder Plot, the unrest that led to the English Civil War. He was living in Paris because he didn't want to get caught up in the fighting. He was a royalist, to be sure, but his main focus was his philosophy and the book he was writing.

Very simply put, he believed that there are two ways that we come to know things. There are some things we know because they are obviously true—that a triangle has three sides, for example. No one has an opinion about how many sides a triangle has. But people do have opinions about things that aren't obviously true, and that's where the trouble starts. The conclusions we come to about the not-obviously-true things, he believed, are governed by how much pleasure or pain those conclusions make us feel rather than any rational thought. He thought this was why people become so dogmatic in their beliefs that they come to blows. The passions, he thought, are the cause of all wars. The solution was for the people to appoint and support a final arbiter—a person who would have the last word, whose decision was final. Hedging his bets, he called that arbiter a monarch, but he meant king.

Leviathan was a book more than a little influenced by the politics of its time. Like Plato before him, Hobbes drew insights from the horrors of war, which fed his intellect and opened his mind. He, like so many others, realized that Aristotle was wrong. He had some better ideas, and he wanted to get them down on paper. By 1651, he'd managed to complete and publish his work, and since then, *Leviathan* has been studied, analyzed, prodded, poked, and turned inside out. But to me, it's about emotions—what they can do and how best to control them. Or, rather, what happens when you can't.

A good argument can be made that much of the Enlightenment was about people responding to Hobbes. One set of responses came from people who wanted a better solution to conflicting opinions than having a single, unchallengeable authority make a ruling. Hobbes argued that you had to be aware of passions before you knew you had them, and that meant thinking about them. But what if opinions could be analyzed without the passions getting in the way? What if, just as Descartes had separated the body and soul into two entities, thinking and feeling could be divided? It would take another 150 years before someone figured out a way around that problem.

Brown's Enlightenment

In the second decade of the nineteenth century, a thirtysomething Scottish man named Thomas Brown stood waiting to give one of his many lectures. His colleague, the ever-popular Dugald Stewart, was unwell once again, and so it had fallen to Brown to step into his formidable shoes. Brown wasn't too worried. Like many of his contemporaries, he was something of a jack-of-all-trades. Brown had studied law, medicine, moral philosophy, and metaphysics. He'd even worked as a doctor for a while before taking up the post as Dugald's colleague at the University of Edinburgh. Brown knew what he was talking about. Besides, he was confident and good-looking, and he had a talent for public speaking and an incredible propensity for coming up with new ways of looking at things. Being so gifted meant that time and time again, he had his students eating out of his hand.

Though he had studied the great philosophers, Brown had been born a few decades too late to personally experience the Enlightenment. But he had taken to heart the one thing its most influential intellectuals managed to agree on: reason is useful, and feelings aren't. Keeping feelings in check had always been important to Western thought. When feelings caused wars—and they were usually the ultimate culprit—it led to nothing but sorrow and misery. When reason caused conflicts, it was to end sorrow and misery—theoretically, at least. The problem was that feelings were still linked to thinking in the philosophy of the day. For one of Descartes's emotions to become a passion or a sentiment, it had to be thought about by the mind. The same was true of Hobbes's passions. That meant feelings always influenced thought, at least on some level. Empiricism, and the modern scientific practice of testing ideas outside an individual's brain using experiments, arose out of a desperate need to disentangle feelings from thinking. But could feelings be removed from the academic table entirely?

Brown, of course, knew about empiricism. He was, in fact, an empiricist himself. He also knew of Hobbes, because no one could claim he was a moral philosopher in the eighteenth century without having read *Leviathan*. As an ex-physician, he also knew that the medical profession understood emotions as external bodily changes and movements that indicated which passions a person was feeling. But there was one philosopher in particular—one of the great empiricists of the Enlightenment—who influenced Brown's thought on emotion more than the others, if only because he totally disagreed with him.

Brown loved the works of David Hume not only because he was a fellow Scot but also because he was—some pretty nasty views on race aside—quite simply brilliant. Hume's first work was *A Treatise of Human Nature*, written when he was only twenty-eight years old. In it, he suggested that rather than thinking there's some sort of battle going on as you try not to think about what you feel, you should just assume that anything you think is being controlled, completely, by feelings. As he put it, "Reason is, and ought only to be the slave of the passions, and can never pretend to any other office than to serve and obey them."[8] Hume was probably right about that, if the latest science is to be believed. Hume watered down his views on the passions a little in later works, but he was always adamant that it was impossible to separate thoughts from feelings, no matter how hard you might try. As for the idea of emotion, Hume sometimes used the word *emotions* interchangeably with *passions*, *feelings*, and *affects*, but he tended to use *passions* for the mental part and *emotions* for the physical part.

Brown took hold of Hume's ideas, and all the other ideas about the passions and emotions he had learned, and tried looking at them in a new way. In a flash of inspiration, he realized that Hume was wrong. There is a way to separate thoughts entirely from feelings, and you do that using emotion. "What if," we can imagine Brown musing, "emotions *are* feelings? What if—in the same way that the brain *sees* something even though the eyes aren't in the brain, and the brain *hears* things even though the ears aren't in the brain—the brain just *feels* things, even though the things causing those feelings aren't in the brain? What if all we do is look, touch, taste, smell, hear, remember, or even imagine things, then just *feel* emotions in our minds straightaway, no thinking needed? That means you can get rid of passions and affects and sentiments and all that nonsense and just create a single, thought-free concept called emotion." Or, as he actually put it:

> Perhaps if any definition of [emotions] be possible, they may be defined to be *vivid feelings*, arising immediately from the consideration of objects, perceived, or remembered, or imagined, or from other prior emotions.[9]

In one lecture given on a (probably) damp afternoon in a (possibly) cold University of Edinburgh lecture theater, the new mental concept of emotion was invented.

What Is an Emotion?

Brown's definition of emotion seems to have caught on. All people seemed to need was the idea that when you feel a feeling, that feeling is called an emotion. Many famous and influential people, including Charles Darwin, started to use the word *emotion* as Brown had. Darwin even wrote a book about emotions in 1872: *The Expression of the Emotions in Man and Animals*. According to this book, emotions in both humans and animals begin when the brain and nervous system react to something, causing changes in the body. Darwin believed that some of these changes could be seen as instinctual actions—such as raising an eyebrow in surprise or shaking in fear—and that many are shared by animals and humans. For Darwin, this was further evidence that humans and other animals have some kind of common evolutionary ancestor and that emotions have some sort of biological source.

Despite the shortcomings of their respective theories, Brown and Darwin had broken new ground by describing a physical mechanism for emotions. But there was still no clarity in their or anyone else's work as to what emotions actually are. That is, until two of the great godfathers of modern psychology, William James and Carl Lange, independently hit on an idea in 1884 that would come to be known as the James-Lange theory of emotions.

The respective backgrounds of these two men had a role to play in the origins of their specific theories of emotions. Both James and Lange were born into wealthy families, and both initially studied medicine, but that's about where the similarities end. Lange stayed in Denmark, his country of birth, for most of his life. He studied in Copenhagen and worked as a physician in the local hospitals. From the beginning, he was gifted, flying through medical school and up the ranks of the medical profession. He was interested in anatomy, particularly the spinal cord and its relationship with pain. His crossover into psychology and emotion theory came when he linked the sensations caused by an excess of uric acid to a form of depression.

James, on the other hand, traveled widely, spending time all over Europe and becoming fluent in German and French. He suffered from bouts of depression and found medical school a challenge, often taking breaks to travel and get his head straight. Though he understood, and even taught, medicine, physiology, and biology, his demons drew him toward philosophy and, in particular, the growing field of psychology. James was a big fan of Brown, and, of course, both men had read

Darwin—few scientists in the late 1800s hadn't. James and Lange arrived at the same conclusion about emotions within a year of each other via different paths, one through research, the other through his own experience. Although there are some technical differences between what they thought, what's important are the general areas where they agreed.

James and Lange's idea was actually quite simple. While Darwin and his contemporaries thought that emotions start in the brain and cause changes in the body, James and Lange thought the reverse— that emotions begin in the body as it reacts to something even before the brain knows about it. Or, to let James speak for himself:

> Common sense says, we lose our fortune, are sorry and weep; we meet a bear, are frightened and run; we are insulted by a rival, are angry and strike. The hypothesis here to be defended says that this order of sequence is incorrect.... We feel sorry because we cry, angry because we strike, afraid because we tremble.[10]

To James and Lange, the body and brain perceiving something and reacting to it happens at almost the same time, and it happens subconsciously. That is the emotion. Only afterward does the brain try to give a name to the emotion from the data it gathers. This bit doesn't require any thought, because the brain has no choice about which emotion you're experiencing. For example, have you ever noticed that your cheeks are warm and bright red and subsequently realized it's because you're embarrassed? That's what James and Lange were talking about. Your body reacts to a situation, and then afterward, you realize what it means. But you have no choice: your body tells you that you were embarrassed. It did not tell you that you were frightened or happy, sad or horny—just embarrassed. Although, of course, you could have been horny *and* embarrassed.

James, in particular, had a massive audience for his paper about this idea—"What Is an Emotion?" At the time, psychology had only recently become a serious field in academia, particularly in his native America. When he published the paper, in 1884, psychology departments and new psychology journals were popping up all over America and Europe. He had a captivated and growing audience, and that audience was dying to be critical of something. Because James's theory is far from perfect for all sorts of tiny reasons I won't get into here, it was attacked, dissected, restated, repudiated, appropriated, and pilloried. But in a way that happens all too often in most fields of science, the fact that it was the starting point for every discussion in the field gave it legitimacy. His understanding of emotions was mentioned repeatedly. Because of this, the English concept of emotions replaced earlier ideas about passions, affects, sentiments, and so on—once and for all.

Coming Over All Emotional

The story of the birth of emotions is one of surprisingly significant impact. Hobbes's work, for example, made a huge splash. His belief that people squabble because they come to different points of view, and that it all boils down to a person's feelings, is partly why most countries in the world have a legal system with some sort of final arbiter to this day, be it the Supreme Court in the United States, the Court of Justice of the European Union, or the Saikō Saibansho in Japan. Hobbes pioneered the idea that positive emotions feel good and negative emotions feel bad, something seen as all but obvious today. He also became the main foil for the philosophers of the Enlightenment. Thinkers clawing through his ideas to find challenges and truths played no small role in the formation of modern democracies, justice systems, and laws. All this in no small part thanks to one man and his ideas about feelings.

As for Brown's new concept of emotions, regardless of your language, if you reach for a psychology or psychiatry textbook, it's the modern English understanding of emotion that will dominate the pages—at least in those books dedicated to the subject. That's because most of the work in the field has either been done by English-speaking academics or, more frequently, been published in English-language journals. If you are a native English speaker, I wouldn't be surprised if it never crossed your mind that emotions are little more than a box that scientists have put some feelings into while leaving out others. Not until you started reading this book, anyway. In fact, the word *emotion* has become so ingrained into the language and culture that we give the word no more thought than we do the word *arm*. The difference is, last time I checked, that I knew exactly what an arm is. By contrast, the last time someone tallied them up (back in 1981), there were 101 different definitions of *emotion* in use by psychologists.[11] Things have only gotten more complicated since.

Every time you say that someone is emotional, or that you need to improve your emotional health, or even that you want to listen to some emo music, you can trace it all back, I believe, to Thomas Brown. With one lecture, the new concept of emotions was invented, and the trench between thought and feelings became more profound than ever before.

But Europe isn't the only place in the world, and, as we'll see in the next two chapters, there are many places where no one had ever heard of Hobbes and Descartes, let alone Brown. People certainly weren't using this new word *emotions*. That was a Western idea that meant little to the African Asante people or the residents of Japan in the nineteenth century.

Nine

A Cherry-Blossomed Shame

Brown's new ideas about emotion eventually caught fire and spread around the world. But this didn't happen instantaneously, and the fire didn't spread everywhere. At the end of the nineteenth century, most of the non-English-speaking world hadn't a clue about this new concept called emotions. This was certainly true of a twenty-nine-year-old Japanese man named Yoshida Shōin as he sat in a prison cell in 1859. He had bigger things on his mind, such as his failed attempt to assassinate a representative of the shogunate and thus prevent the signature of a treaty between the United States and Japan.

Shōin was something of a troublemaker. He'd been stripped of his samurai status for studying *rangaku*—Dutch studies—thereby polluting his mind with the ideas of foreigners. Overseas travel had often interested him. He then committed an even greater taboo in an attempt to learn more foreign philosophy by seeking passage on an American ship that had anchored off the shore of Japan. He was refused by the captain, and his attempt to go aboard got him placed under house arrest, then imprisoned. He was eventually released and stayed out of trouble for a bit. But getting into the assassination business got him locked up again, and this time it was more serious.

Shōin didn't know it—technically, he would never know it—but he was to become a hero. He wasn't your typical hero, however. His friend Taizo Masaki described him to Robert Louis Stevenson as

> ugly and laughably disfigured with the small-pox.... His personal habits were even sluttish. His clothes were wretched; when he ate or washed he wiped his hands upon his sleeves; and as his hair was not tied more than once in the two months, it was often disgusting to behold. With such a picture, it is easy to believe that he never married. A good teacher, gentle in act, although violent and abusive in speech, his lessons were apt to go over the heads of his scholars, and to leave them gaping, or more often laughing.[1]

And bear in mind, this was coming from a man who cared for and respected him.

Born in 1830 to a modest-ranking samurai, Shōin was adopted, as was the custom, by another samurai—Yoshida Daisuke—who died soon after, leaving Shōin the heir to the Yoshida line at just five years old. He committed his first serious act of defiance by going on a walkabout across northern Japan before he had permission to do so. He was twenty-one and eager to see as much of the world as he could. On his return, he was stripped of his samurai status and given a punishment that, as far as he was concerned, was more like a reward: he was told he had ten years to study anywhere in Japan he wanted. It was during his return home from the first year of study that the Americans showed up.

In 1853, Commodore Matthew Perry of the US Navy swung four ironclad "black ships" into Edo Bay and demanded, on behalf of his government, that Japan open trade. While Japan had remained peaceful and isolationist, it had also lagged behind in technological advances. It was no match for the Americans. Following this incident, Shōin—along with some others—became concerned about Japan's relative inability to defend itself from outsiders. Civil peace was all well and good, but it meant that there was no well-trained fighting force equipped with the latest and greatest weaponry.

The seeds of this problem were planted in the early days of the Tokugawa shogunate. Ever since 1603, the Tokugawa clan had ruled Japan. On the one hand, they'd brought peace, ending the constant warfare of the feudal samurai. Soon after coming to power, the Tokugawa split the country into 260 smaller, more manageable domains. Each was given to a *daimyo*, or lord, who ruled his little section almost independently. The once great *katana*-wielding samurai, whose battle prowess has become the stuff of legend, swapped their fighting swords for writing brushes and became the administrators for their *daimyos*. This kept local and national politics separate. Higher-ranked samurais and their *daimyos* focused on local matters, while the shogun and his advisers tackled things at a national level. It kept things peaceful.

On the other hand, the Tokugawa cut Japan off almost entirely from the outside world, save for a small Dutch colony at Nagasaki. The Tokugawa shogunate wasn't interested in what was happening beyond its shores. It was fed up with being seen as an appendage of China—just a small cherry-blossom-colored group of islands off the coast of a seat of great wealth and power. The shogunate banned foreign philosophy and the sharing of foreign ideas completely. This didn't just mean studies that were Dutch, despite the name *rangaku*. It meant anything not Japanese—Christianity, Islam, Buddhism, Confucianism, Daoism. It meant any tradition—intellectual, spiritual, artistic—that didn't have roots in the era before feudal Japan, back when the Shinto religion and its many gods dominated thought and belief.

The nation was at peace, and it was fiercely independent. But the arrival of the Americans changed all that. Japan had to be able to defend itself. Stability could no longer be guaranteed by isolation. Something had to be done. A spark had to be rekindled in the samurais' warrior souls.

Shōin decided to try to ignite that spark using the Japanese concept of *haji*, or what we might today call shame.[2] To Shōin, shame wasn't some internal feeling you keep to yourself: it was a social emotion, something shared by a group. When he carried out his assassination attempt, he was motivated by a belief that it was worse to feel the shame of inaction than to commit a crime in the service of changing an unacceptable situation. Getting involved in national politics was a crime, he acknowledged, but to not do so would cause *haji*, and that was worse. If all the samurai remained shameful in this way, it would mean calamity for the nation of Japan.

It's easy enough to translate *haji* as "shame," but as I'm sure you've come to expect by now, the meanings of the two words don't perfectly match up. *Haji* is a particular type of collective shame, and in order to understand it and its importance, we need to know a bit about the melting pot of traditions that informed the way Japanese people at the time understood it. But before we do, it's worth taking a look at what academia has to say about shame, because quite a bit of modern research into that emotion can help us understand the emotional regime that dominated Japan at the time of Shōin.

What Is Shame?

Scientifically speaking, shame occurs when our parasympathetic nervous systems are activated by *attunement disruption*—that's when a person falls out of sync with the emotions of the people around him or her. Or, to put it another way, it's activated when you realize that you've broken the rules of an emotional regime. Its role seems to be communal; that is, it's a social emotion that lets us know when we've crossed a moral boundary.[3]

Shame has been linked to something called *self-discrepancy theory*, an idea developed by Edward Tory Higgins in 1987. He suggested that we all have an internalized *ideal self*. You might see yourself as moral, intelligent, and upstanding, for example. And why shouldn't you? You drive carefully, say "please" and "thank you," and are respectful at all times—a right and proper member of society. To become our ideal selves, there are ways we ought to behave, and the self that behaves in those ways is called the *ought self.* The ideal self is something we aspire to be, who we hope we are, ideally. The ought self is the self that has a sense of duty, sticks to codes, and fulfills obligations. It does the things one ought to do. Your ideal self drives carefully because it should and because not doing so would put other people in danger, which is morally unacceptable. Your ought self drives carefully because it's the law, and you're not the sort of person who breaks the law. But there are always gaps between this ideal/ought self and the *actual self*—for example, it might be the case that, regardless of your self-perception, you are actually a terrible driver who swears like a sailor. When you realize that your actual self is not living up to your ideal/ought self, you might feel the sensation of shame.[4] Your mood changes; your muscles move involuntarily, especially in the face and shoulders as they appear to droop; and your endocrine system releases a cocktail of stress hormones—cortisol, adrenaline, and just a smidge of oxytocin. It's that oxytocin that seems to make shame social, reinforcing the bonds we've broken.

Then shame does something else—it initiates a fight-flight-freeze reaction: fear. Usually, we'll freeze, worried about what people think of us. But if we believe that our shame has been caused by someone else, the need to fight can take over. We get angry. Still, as a social emotion, shame isn't just about individual feelings. According to one anthropologist—Ruth Benedict—the balancing of shame, fear, and anger can underpin the emotional regimes of entire cultures.

A Shame Culture?

In a book published in 1946 called *The Chrysanthemum and the Sword,* Benedict described Japan as a "shame culture." She was contrasting

Japan with America, which she called a "guilt culture." The main differences, according to her, are that guilt cultures focus on people feeling guilty about things they've done. People in guilt cultures worry about personal punishments—prison, hell, and so on. People in shame cultures such as Japan, she opined, are more worried about being outcasts from society, about bringing dishonor to the people around them. She also identified another type of culture—fear culture. This is one in which everyone is terrified to do wrong on pain of harm or death. The cultures that produced the witch crazes we explored above would be, in Benedict's view, fear cultures.

One of the biggest criticisms leveled against Benedict is that she focused not on the Japanese people as a whole but solely on the country's elites. In addition, it's important to note that Benedict wrote about Japan during a time of national upheaval; that is, at a time when Japanese culture was undergoing rapid and dramatic changes. For one, Japan had begun starting wars of aggression rather than keeping to itself. But—and this might seem a little odd, since she was an outsider—Benedict's ideas were embraced by quite a lot of Japanese people. Since its translation, *The Chrysanthemum and the Sword* has sold two million copies in Japan. It spearheaded an uptick in the publication of *nihonjinron*—works about "Japaneseness." It even embedded the notions of shame culture and guilt culture into the national psyche.

Of course, the reality is that all cultures indulge in a bit of shame, a bit of guilt, and a bit of fear. They also are defined by a bit of love, a bit of desire, and a bit of bravery. Calling Japan a shame culture might seem a bit reductive—and it almost certainly is—but the Japanese concept of shame, *haji*, did play a role in bringing the country back into the world. To understand *haji*, though, you need more than a modern scientific understanding of shame. *Haji*, at the end of the Tokugawa shogunate, despite its attempts to purge Japan of foreign ideas, was based on a constellation of beliefs about shame that had come to Japan over the previous millennium, starting with the ancient practice of Shinto.

The Shame of the Gods

According to the Shinto religion, at some point in the distant past, the first of the gods, Kuninotokotachi-no-Kami and Ame-no-Minakanushi, created two divine beings, Izanami and Izanagi. They were given the job of making the first land. They used a heavenly spear to churn the sea below the bridge that linked the heavens and the earth, drawing land out of the sea to create the island of Onogoro-shima. Happy with what they'd done, they decided they would live on this new island. They built a pillar of heaven to support a great palace, and they began to raise god-children—a lot of them. They also created the remaining islands of the Japanese archipelago for humans to live on.

Eventually, tragedy struck. According to one of the oldest Japanese imperial texts—the Kojiki (712 CE)—Izanami died giving birth. Izanagi was so furious and heartbroken that he killed the child who he thought had murdered his wife. Don't worry too much, though. Because the child was really a god-child, his murder created dozens of deities who were scattered throughout heaven and earth. Izanagi's sorrow then took him on a journey to Yomi, the underworld, to find his lost love. He wasn't supposed to go there. Once he found her, he pleaded with her to return, but she couldn't. She had eaten the food of the dead and was no longer able to exist among the living. But Izanagi didn't want to leave his love. He wanted to see her beautiful face again. That, too, was forbidden in the darkness of the underworld.

One night while Izanami slept, Izanagi decided to use his long hair to make a torch so he could light up the gloom and gaze upon his wife.

As the light struck her body, he didn't see the exquisite form of his spouse but rather a twisted corpse, her flesh rotting and maggot-ridden. Izanagi screamed and fled in horror, rushing for the cavern door that led to the world of the living. His wife, filled with shame and rage at what he'd done, chased after him, desperate to stop him from leaving. Izanagi leaped through the entrance and rolled a boulder over it. The worlds of the living and the dead would be separated forever. His furious wife, still consumed by the *haji* she felt for having been seen by her husband, cursed the living, bringing death into the human world.

Izanagi performed a ritual cleansing known as *harae* to rid himself of the defilement of what he had seen. According to the Kojiki, that defilement was also *haji*. Just as his wife had felt shame at being seen, Izanagi was ashamed of himself for breaking taboos. He had been misbehaving, wanting something he shouldn't have. The cleansing practice of *harae* remains a Shinto ritual to this day.

This ancient Japanese myth tells us something about how eighth-century Japan understood emotions and how neatly they seem to fit into modern scientific ideas about shame. Eighth-century Japanese emotions were based on how you *ought* to feel and behave in a given situation. Grief over the loss of a wife is appropriate, but breaking taboos for your own gain is not. To do such a thing would be to move away from your ideal self, to break the rules of the emotional regime. It would, or at least should, cause *haji*. At the same time, eighth-century *haji* brings with it elements of disgust and a need to clean the shame away—a need so powerful that it's still practiced today.

In a fantastic example of the melting pot that is Japanese culture, the roots of this idea about feelings come not from the Shinto religion but from Confucianism. To explain this influence on *haji*, let me tell you another story, this time adapted from the writing of the great Confucian philosopher Mencius.

The Well of Qing

One evening sometime in the third century BCE, a woman was on her way to fetch water from her local well. The well she was walking toward was deep and plentiful, but what made it special also made it dangerous. All too often, and to her frustration, this unguarded hole had children playing around it. As she got near, she saw one child getting ever closer to the rim of the well. Then the child disappeared. The woman instantly dropped her buckets and ran toward the pit. Peering in, she could just make out the body of the child floating motionless in the water.

Her immediate feeling was alarm and distress. Her concern for others overwhelmed her. This is because her most important asset, her *jen*, or goodness, overcame her. She was also filled with a sense of *yi*, the desire to do good, be moral. And that's why she didn't just run around waving her arms in panic. It's why she didn't run away, fearful that she might be blamed. It's also why she drew on the five virtues—kindness, moral righteousness, decorum, wisdom, and sincerity—and began to look for help. She knew that the best way to help the child was to act as she had been taught, to follow the *dao*, or guiding principles, that led her through *li*, the right conduct, rituals, and behaviors. More important, she had to keep the appropriate feelings at the proper levels. She was not to be callous or panic. This process of controlling your feelings involved the *zhongyong*, often translated as the "doctrine of the mean." This doctrine was so important that it had its own book dedicated to it, a book that told her:

> While there are no stirrings of pleasure, anger, sorrow, or joy, the mind may be said to be in the state of Equilibrium. When those feelings have been stirred, and they act in their due degree, there ensues what may be called the state of Harmony. This Equilibrium is the great root from which grow all the human

actings in the world, and this Harmony is the universal path which they all should pursue.[5]

The woman sought help appropriately, as she had been taught; she had been a good student. She had a choice of—depending on which books she read—four, six, or seven emotions to draw from. The basic four—from the *zhongyong*—were delight (or love), anger, sorrow, and joy. These were innate feelings the author believed all humans were born with. If she had read the work of the third-century BCE philosopher Xunzi, she might have also known about the two additional learned passions, like and dislike. If she'd read the rest of the collection that the *zhongyong* comes from—the *Li Ji*, or *Book of Rites*—she would have also known how best to act when feeling fear. Whatever she felt, she knew to keep questioning, disciplining, and educating herself in order to be a better person in society. She knew anger at the drowned child's parents was of no use and that forgiveness for their actions and sincerity about what had happened was what mattered now. She knew fear and sorrow, though appropriate, shouldn't be overindulged. She also understood her *hsiao*, her deep reverence for and relationship with the people of her village. It was second only to the *hsiao* for her family. It was more important than her dislike for the wild nature of the child.

She knew all this in the blink of an eye. She understood her *li* so well that she immediately rushed to the family of the child. She told them, calmly but with urgency, to bring ropes and ladders. Because this was the way of Confucius. The way of balance.

This might click together even more tightly in your mind if I point out that the ancient Chinese word *qing*—the closest equivalent to the English word *emotions*—is better translated as "a practice of wants and desires embedded in the real world." *Qing* should always be balanced; never too much, never too little. The route to that balance is through *li*—learning your society's proper rituals, customs, and ways to behave, even when they go against your personal wants and desires.

That, from the Confucian point of view, is why Izanagi felt shame. The most profound sense of shame came from failing to act in accordance with *li* and diverging from the *dao*. It doesn't matter if you're a peasant farmer or a god: to contradict your society's *li* is to bring shame on yourself. *Li* isn't the same as *dharma*. It isn't a path or a destiny you are supposed to follow regardless of whom you might hurt or how hurt you may feel. It's a way of achieving balance in society by balancing yourself. Destiny has nothing to do with it.

Let us now return to Japan. In the Tokugawa period, one of the things the shogunate wanted to do was resurrect older Japanese Shinto beliefs through *kokugaku*, or "native studies." This was not only a rejection of Dutch studies; it was also intended to push back against the Confucian influence on Japanese thought. And yet during the Tokugawa period there was a resurgence in Confucianism, known as *Shushi-gaku*, or neo-Confucianism. The Japanese translated *li* to *ri*, and the neo-Confucians linked *ri* to a concept they understood: *zhong*, derived from the aforementioned *zhongyong*. *Zhong* is loyalty to or a filial reverence people have for their family, friends, and, most important, rulers. It helped strengthen the social structures the shogunate had put in place. It also created a more in-depth focus on Japanese culture as the source of *ri*.

But *Shushi-gaku* was also entirely at odds with the other dominant belief system in Japan up to that time, one that represents another link in the chain of ideas that forms eighteenth-century *haji*—Buddhism. The neo-Confucians absolutely believed that the world was real, physical, and tangible and that actions within it had consequences. The Buddhists didn't think anything was real. Not you, not me, not this book. The world was an illusion, and *ri* and *zhong*, to borrow a line from Douglas Adams, were doubly so.

How to Feel Nothing

Initially, the Tokugawa shogunate tried to repress the foreign influence of Buddhism.[6] But by the time the nineteenth century rolled around, things had improved so much for Buddhism that one of the Tokugawa shoguns even supported the printing of the entire Buddhist canon.[7]

The strain of Buddhism that swept across Japan in the fifth century and continued through the Tokugawa shogunate wasn't quite the same Buddhism I described in relation to Ashoka. This newer form was based on a relatively broad spectrum of Buddhist traditions and belief systems usually blanketed under the term Mahayana, or "great vehicle." (Some scholars argue that a more accurate translation is "great understanding.") It shared a lot of its ground rules with previous strains of Buddhism, especially when it came to feelings, but there were a few notable differences. One is the idea of a *bodhisattva*—that is, a person striving for Buddhahood. In many sects of Buddhism, you can't be a *bodhisattva* unless it's been foreseen by someone who has achieved Buddhahood. In Mahayana Buddhism, anyone who begins on that path is a *bodhisattva*, whether a current Buddha has foreseen it or not. More to the point, Mahayana Buddhists believe that anyone who accepts their canon of texts, the Mahayana *sutras*, will at some point head on the path of a *bodhisattva*.

Mahayana Buddhism was also a bit more mystical than the older types of Buddhism. In this tradition, Buddhahood was not just a path to *nirvana* but also a path to becoming a transcendental, immortal being who has risen beyond the confines of this world and can assist other beings. They believed that the Buddha himself, Siddhartha Gautama, acted as a human despite being a spiritual king. He had come to our earth briefly to guide humanity.

Another important aspect of Mahayana Buddhism is the doctrine of *sunyata*—a word best translated as "emptiness" or "voidness." The exact meaning of this concept shifts slightly among the various threads, or sects, of Mahayana Buddhism, but it essentially refers to an awareness that everything is empty at its core—that life, the world, even *dharmas* are a dream (*svapna*) or an illusion (*maya*). It means that when you cling to desires, you're actually nothing, clinging to nothing. *Sunyata* can also refer to the notion of a Buddha-nature or the ability to become like the Buddha that lies within us. Some types of Mahayana Buddhism even go so far as to say that *sunyata* is *nirvana*.

Japanese Buddhists tended to belong to one of three threads that appeared within Mahayana Buddhism. The first, Nichiren Buddhism, was introduced to Japan by a man named Nichiren, who lived from 1222 to 1282. He thought everyone could achieve *sunyata* and was endowed with a Buddha-nature. The trick to tapping into it was repeatedly chanting the mantra *Namu myōhō renge kyō* (meaning something along the lines of "Glory to the *dharma* of the Lotus *sutra*"). The *Myōhō renge kyō*, or Lotus *sutra*, is thought by many Buddhists to be the final word in Buddhism—in fact, just saying its name can help you on the path. But more critical for us is Zen Buddhism, because by the nineteenth century, when our friend Shōin was in prison, it was more prominent.

Zen Buddhists believe that everything is part of Buddha-nature. Rather than the capacity to reach the emptiness of *nirvana* and become Buddhas, Zen Buddha-nature is the ability to understand that everything is connected and that everything changes, constantly, from the brain in your head to the rocks on a distant exoplanet. Interestingly, Zen *sunyata* is not a void or nothingness.[8] Instead, reality as we understand it is an illusion, and that includes feelings. But there is a reality beyond that, and that's what people, ideally, ought to strive for.

Zen Buddhists don't think the way to achieve *nirvana* is through learning *sutras* or chanting about them. They believe you have to meditate, clear the mind. They aren't the only Buddhists who meditate, of course; meditation has been part of the religion ever since Siddhartha

Gautama sat under the Bodhi Tree. But they reason that because the Buddha himself didn't need any pomp, ceremony, *sutras*, or slogans to achieve *nirvana*, neither does anyone else. After all, those rules clutter a mind that's supposed to be empty.

Zen Buddhists use emotions as part of their meditative process. To Japanese Buddhists in the nineteenth century, as today, the goal was *mono no aware*: "the pathos of things" or "the deep feeling of things," a kind of pity. This means being aware of the feelings you have and realizing that your feelings match the *hon'i*, or essential nature, of where you are and when you are. Let me explain by sharing a poem.

> *Loneliness:*
> *The feeling came*
> *From nowhere I can name;*
> *On an evergreen covered mountain*
> *On an autumn evening.*⁹

This poem was composed by the Japanese Buddhist monk Jakuren, who died in 1202. Here, Jakuren is dissolving his personal loneliness into the world around him. But he isn't really alone; he is surrounded by the beauty of the mountains and the setting sun. *Mono no aware* is a path to *nirvana* that doesn't require the expulsion of all feelings: instead, you understand how those feelings relate to the world around you. You use them to dissolve into the emptiness of the universe. Become one with it. Realize how the way you feel interacts with the illusion that is the world.

But what does that have to do with *haji*? In the case of Japanese Buddhism, *haji* occurs when people see you in a way you'd rather they didn't, when their knowledge of something you'd rather keep hidden makes it impossible for you to dissolve neatly into the void. You might call this a sort of *attunement disruption*.

A Japanese form of dance-drama known as Noh might help to

illustrate the point. In one such drama, called *Eguchi*, two characters, a monk and an *asobi*—a member of a group of wandering dancers once associated with prostitution but later a quasi-order of Buddhist nuns— are speaking to each other. At this time in Japanese history, sex work was thought of as a source of great *haji*. The monk recognizes the *asobi* and remembers when she was a sex worker. He recites a poem asking her for "lodgings" and "refuge." (I'm sure it's not too hard to work out what "lodgings" and "refuge" mean in this context. He wasn't just asking for a bed to sleep in overnight.) She, of course, refuses and is overcome with *haji*.[10] Being reminded of who she once was is a source of shame for the *asobi*. More powerfully, it's the *haji* born from the monk's ability to remind her of a past *actual self* that brought feelings of shame when contrasted with her *ideal self*, which she was striving toward.

Historian Gary Ebersole has pointed out that the *asobi*'s *haji* is a lot like the shame Izanami felt when her husband saw her corpse: "The emotion of shame is produced in a character when someone views her in a way that does not match her own sense of identity or her public self-presentation."[11] You become fearful that someone knows something about you that you want to keep hidden—for example, that you are decomposing or used to work in a profession thought of as taboo. According to linguist Gian Marco Farese, the fear of people finding out something you'd rather they didn't is a central part of *haji*, even today.[12] In a religion whose goal is to evaporate out of a world thought of as an illusion, *haji* had the power to make everything feel mighty real.

Revolutionary Shame

Now we can jump back to 1859 and reunite with our old friend Shōin and the samurai. The samurai at the time had their own sense of *haji*, distilled from the many influences they'd absorbed during their lives. It included the Buddhist reluctance to reveal something they'd rather keep secret, the Confucian ideal of emotional restraint, and the disgust—often self-disgust—associated with Shinto notions of *haji*. The samurai had a culture of honor that revolved around a constellation of ideas: *na*, or "name"; *iji*, or "pride"; *mengoku*, or "face" (as in "saving face"); *meiyo*, or honor; and two types of shame, *chijoku*, meaning stigma or disgrace, and *haji*.

As I've said, *haji* was a major element of the Japanese emotional regime and one that Shōin wanted to tap into. He believed it was the key to rekindling long-dormant samurai honor.[13] This particular *haji*, it's worth noting, was primarily a concern of the elite. The average Japanese farmer was unlikely to have been quite as worried about honor and shame as the samurai who supervised him. This is one of the issues with Ruth Benedict's book, of course. She reported shame in all of Japan when only looking at part of it. But unlike Benedict, Shōin was targeting the elite. He wanted to rile up his superiors, give them a shake, make them understand their duty to the world around them. Surely finding yourselves unprepared when a foreign power arrives with weapons impossibly superior to your own would be a way to ignite *haji* in the powers that be. He believed that Japan had been forced to reveal something about itself that it would rather other countries didn't know: it was a nation forced to come to terms with the way its *actual self* fell short of its *ideal self*. A nation that had failed to behave as it ought to, that had strayed from its *ri*, placing itself in danger. A nation that, Shōin thought, ought to be disgusted with itself.

It didn't end well for Shōin. On November 21, 1859, after confessing to the assassination plot, he was executed. The last words he wrote were about his parents: "Parental love exceeds the love for his parents. How will they take the tidings of today?" We will never know. What we do know is how fifty or so of his students took it. They threw themselves into overthrowing the shogunate and reforming the nation. The shame of not being ready for outsiders, of letting Japanese society decline, formed the bedrock of what's now known as the Meiji Restoration.

Once the Tokugawa were overthrown, many of the new rulers were Shōin's students. One notable individual was Katsura Kogorō, who became known as one of the *Ishin no Sanketsu*, or Three Great Nobles of the Restoration. The new government used the shame of Japan falling behind as incentive to industrialize Japan at breakneck speed, to modernize, to catch up—eventually, to surpass. That scruffy young man who had lost his rank as samurai set something in motion, something I suspect he was happy to die for.

Ten

The Rage of an African Queen

Before we head off into the scientific understandings of emotion that dominated the minds and machinations of the twentieth-century West, I want to stay away from Europe for a while. I want to visit a part of the world where another emotional regime, one different from the ideas set forth by Brown and his followers, dominated — where another way of understanding what feelings are flourished and still flourishes. Let's turn now to the glimmering coast of West Africa, where at the turn of the twentieth century an Akan queen was getting ready to do battle.

Yaa Asantewaa, the *edwesohemaa*, or queen mother of the Edweso tribe of the Asante, was furious. For centuries her people traded with, and fought both for and against, the British. They once had their own great empire, an empire whose expansion made them rich through the sale of gold and, sadly, enslaved criminals and prisoners of war. They matched the British in battle throughout the eighteenth and nineteenth centuries, becoming one of the few peoples on the African continent to successfully fight off repeated colonization attempts. In 1896, however, the British finally managed to take control of the region, then known as part of the Gold Coast and now known as southern Ghana. Deporting the Asante chiefs and replacing them with puppet leaders was the next step in the usual method of conquest. The British did this, but they missed a critical seat of power—a quite literal seat of power. A seat that the British governor of the region, Frederick Mitchell Hodgson, badly wanted.

The Golden Stool with its immediate caretaker, January 31, 1935 (*National Archives, CO 1069-44-6, Colonial Office Photographic Collection, Africa Through a Lens project*)

On March 28, 1900, Hodgson summoned the new, inexperienced Asante chiefs together. He mistakenly thought they were disorganized and incapable of fighting back, regardless of his demands. Hodgson wanted something precious of theirs to consolidate British rule in the region, something that he knew kept the spark of rebellion alive. It also just so happened that this something was rumored to be made out of solid gold. According to Methodist missionary Edwin Smith, who might or might not have heard this firsthand, Hodgson demanded that the Asante give him the seat of their power, both figuratively and literally—the Golden Stool.[1]

Once again, the British were after something shiny and precious that wasn't theirs. This enraged Yaa Asantewaa, but her fury had nothing to do with the stool's financial value.

The *Sika Dwa Kofi*, meaning "the Golden Stool born on a Friday," was the primary seat of Asante power. The tribe believed it came down from the sky one Friday and landed in the lap of the first Asante king, Osei Tutu. The Asante also believed that any stool you sat on housed your soul while you sat on it. Each element of the Asante leadership had its own stool. It was believed that once you sat on the stool, your soul would be connected to the souls of the people you represented. For example, should you be the occupant of a stool that put you in charge of the tribe's warriors, your soul and the souls of all the warriors would be connected via the stool. (In fact, to this day, positions of power within Asante culture are called stools.) The Golden Stool had particular significance. Its emotional ties were deeper. The Golden Stool united the soul of the entire nation through the ruler who sat upon it. Hodgson's outrageous desire to steal it did not go down well with the queen mother.

Yaa Asantewaa was a powerful woman. The Edweso tribe had emerged as the ruling clan after years of vicious civil war. Because of this, Yaa Asantewaa wasn't just the mother of a chief. She was the mother of the exiled king of the Asante. She was *asantehemaa* as well as *edwesohemaa*—the queen mother of all the Asante. Every position of power in the Asante territory was occupied by a pair of male and female leaders who oversaw the affairs of men and women respectively. Each was given a stool to represent his or her status. Yaa Asantewaa had an influential stool in the ruling Kotoko Council, granting her the power to distribute land, get involved in the legal process, and wage war. At that moment, her son was unable to occupy it, so following tradition, she sat on the Golden Stool.

Yaa Asantewaa wanted to exercise some of her power. She was enraged by what she was seeing. Not only had Hodgson demanded her culture's most sacred object, the chiefs in the council also seemed to be considering acquiescing, using words such as *compromise* and *offer money* in their deliberations. They bickered. They argued. The queen mother had had enough. To her, the path was clear. She stood up and yelled at the men around her.

How can a proud and brave people like the Asante sit back and look while whitemen took away their king and chiefs, and humiliated them with a demand for the Golden Stool. The Golden Stool only means money to the whitemen; they have searched and dug everywhere for it. I shall not pay one *predwan* to the governor. If you, the chiefs of Asante, are going to behave like cowards and not fight, you should exchange your loincloths for my undergarments.[2]

Then she snatched a rifle and fired it in front of the gathered men. This was more than a rebuke. It was a reminder that Asante women were just as capable of fighting as men. If the men wouldn't defend the Golden Stool, the women would. It was the first shot of rebellion, the hammer of resistance striking the land. It was also enough to startle the gathered men into action. They knew Yaa Asantewaa could be difficult, capricious, and belligerent, but they knew she was right. The first course of action was, of course, to get drunk—to spend the whole night drinking in honor of the gods. The second was to vow to follow their queen mother into battle. Yaa Asantewaa led a group of warrior men, and some warrior women, into rebellion against their oppressor.

What drove her to do this? On the surface, it seems obvious. The British wanted to take her people's throne from them, and that made her angry. But that explanation, aside from being rather shallow, only works if we ignore how the Asante, and indeed the bigger group they are part of—the Akan—view the world and understand their feelings. To understand Akan anger, let's look at the way modern science explains anger so we can see how well it does, and doesn't, match the rage of an African queen.

What Is Anger?

Modern psychology claims that we get angry when something happens that makes us feel threatened or that somehow holds us back from something we want. When angry, we might attack the threat or lash out in an attempt to get beyond the frustration. So far, so good, and this fits in nicely with the idea of anger as the Greeks and Aquinas would understand it, at least to some degree. It also works with the Asante, as you'll see. But science doesn't stop there.

Neuroscientifically speaking, anger stimulates the amygdala and several other areas connected with fear. Anger, it seems, can be a reaction to fear. After all, fear is, by definition, a response to a threat. Specifically, a system called the amygdala-hypothalamus-periaqueductal gray seems to control our "reactive aggression," which is, as you might imagine, aggression that occurs as a reaction to something. But emotions are more than just a stimulus and response, and modern neuroscience knows this. The brain's frontal cortex—the bit where plans are made and thinking is done—is also linked to anger. The issue is that the amygdala circuit acts faster than our ability to think, so when someone appears to be lashing out unthinkingly, that's precisely what's going on.[3]

But anger doesn't just work with our old friend fear: it can also be found taking the side of desire. Being frustrated—being held back from getting something you want—can cause anger, too, although the level of that rage seems to depend on upbringing. At one extreme, a lifelong Buddhist might never lose his temper simply because his wishes have been thwarted. At the other extreme, some people feel anger at the slightest frustration, believing themselves entitled to fulfill their desires without any effort at all.

Anger, like shame, is thought of as a social emotion—that is, an emotion that always has a target. When you feel angry, it's because someone or something has made you angry. You feel anger toward that something or someone, even yourself. This is different from an emotion such as sadness, for example: when you feel sad, you are unlikely to feel sadness *toward* the thing that made you sad. You might feel anger toward it, though.

The queen of the Asante had a target. She had a frustration (the wish of the British to take the Golden Stool), and she had an imminent threat (the British). It's no surprise she acted with rage.

In the Lap of the Gods

In some parts of the world, understanding emotions, especially historical emotions, can be more than a little challenging. Whereas people in Europe, North America, and Asia tended to record their thoughts about feelings in writing, people living in places such as the western coast of Africa didn't. The methods we historians of emotions use to study how emotions operated in such cultures are, by necessity, a little more creative than what we normally do—pore over historical documents. There can be no close readings of texts, no trawling through archives, no pondering of the significance of headstone inscriptions. Instead, we borrow from other disciplines, such as linguistics, because language can tell us a great deal about the way people feel and felt. You can learn a tremendous amount about the way a given group understands emotions by tracing the evolution of its language and by identifying words that share similar meanings as well as stems (parts of words) and roots (origins). I touched on this technique in the previous chapter when I used the linguist Gian Marco Farese's definition of *haji*.

You can also find bits of emotional language passed down through generations orally in parables, legends, and proverbs. The Asante, and the Akan more broadly, like their proverbs, which gives us a lot of material to go on. The Akan also have a shared group of dialects that are pretty much understood by everyone who speaks them interchangeably, despite some occasional significant differences. But we'll be drawing primarily on the Asante language Twi, because that's the one Yaa Asantewaa spoke.

What follows is a journey through Akan emotions as understood through language. But I want to lay a little groundwork first, because to understand a language, we first have to understand how that language is used. And for that, we have to know how the Akan understand the world.

The Akan believe that a human's *ōkra* (soul) is the spark of life that comes from Onyame (God). It's not quite the same as the conception of the soul in Abrahamic religions. In Islam and Christianity, the soul is thought to be the bit of us that thinks, wills, imagines, and so on — the conscious part, or the mind. Try to imagine a similar sort of mind but take away the actively thinking bit. It's not as hard as you might believe.[4] Think of all those times you left home and traveled to work, by car or train or whatever, and couldn't remember your journey. You consciously bought a ticket, drove your vehicle, rode your bike. But you weren't *thinking* about it. Your brain obviously knew what it was doing. It was able to drive or ride, follow directions, and get you where you needed to be safely. It just happened subconsciously — below the level of attention. That almost automatic part of your brain is the *ōkra*.

The *sunsum* is a part of the *ōkra* found in all living things. One might call it supernatural or spiritual, but not in the sense that it isn't *real*. Most Christian and Islamic notions of the supernatural spirit think of it as something immaterial, belonging to a realm that rarely interacts with material things. This is not the case with the *sunsum*. You can see the effect of the *sunsum* by the influence it has over the physical world. It's just not perceivable directly. This is because the *sunsum* is, to quote African philosopher Kwame Gyekye, quasi-physical — that is, it's not strictly part of the world that you see or touch or hear. But it can take on the properties of all living things. The *sunsum* is essential for anything in the physical world to act. This is different from the Abrahamic religions, which hold that God imbues a lump of flesh with a soul to get it to work, much as a driver gets behind the wheel of a car. The *sunsum* is the car; it *is* the driver. It's the difference between inanimate matter and life. Each of us—me; you; my cat, Zazzy—has, or, rather, is, a *sunsum*. It's the ability of things to do things—for actions to happen.[5] It is the bit that thinks, desires, and does things.

As far as feelings are concerned, the *sunsum* is an essential part of your adult personality. The word is often used to discuss people's personal dispositions, including how dignified their presence is and how weighty or powerful their personalities are.[6] But complicated feelings, the things we might call emotions, exist outside the *sunsum*. There are certain things the Akan would say about their *ōkra* that they'd never say about their *sunsum*. For example, the *sunsum* can't be sad, worry, or feel the urge to run away, but the *ōkra* can.[7]

Gyekye believes the Akan idea of the person is bipartite—that is, of two parts: the *honam* (body) and the *ōkra/sunsum* (soul). The latter is thought to survive death, the former not. Some Akan seem to believe the body and soul are so close that they can't be separated. If you were raised Muslim or Jewish or Christian, you might think this would mean that the *sunsum* and *ōkra* would die with the *honam*. Not so. In order to understand why, you must let go of the ideas about the soul posited by the world's major religions. To the Akan, the soul is not some incorporeal, immaterial brain beyond the physical realm. Instead, they think of the *ōkra* and *sunsum* as a substance that can't be destroyed.[8] This substance, they believe, is passed on through conception and can be found in the *ti* (head). It means that, at death, there are two substances—the mortal flesh of the *honam* and the immortal substance of the *ōkra* and *sunsum*. Only the *honam* is destroyed. Eventually, the substance that is your *ōkra* and *sunsum* will be reincarnated further down your bloodline.

The *ōkra* is passed from mother to child through the *mogya*, or blood. The *sunsum*, some believe, comes from the *ntoro*, described by Gyekye as "sperm-transmitted characteristics."[9] The *ntoro* isn't just sperm; it's part physical, but it's also what one might call spiritual. The two are the physical means by which part of the soul is bound by a new body. This is the process through which the Akan believe people inherit traits from their parents and ancestors.

The *mogya* serves to link people by their maternal ancestry into one of seven main groups: the Agona (parrot), Aduana (dog), Asenie (bat), Oyoko (falcon/hawk), Asakyiri (vulture), Asona (crow), Bretuo (leopard), and Ekuona (bull). Each of these maternal groups has its own greeting response and ancestral home. They are often grouped around a cluster of towns. Sometimes they have roles to play, such as the Bretuos' role in warfare and the Asenies' function as royals of their region. Yaa Asantewaa was probably Asenie. Or at least she would have assumed that role.

The *ntoro* is where people get what is best described as a temperament or personality. There are twelve types, and they link to paternal ancestral ties. They are:

1. Bosompra (the tough)
2. Bosomtwe (the empathetic)
3. Bosomakom (the fanatic)
4. Bosompo or Bosomnketia (the brave)
5. Bosommuru (the respectable)
6. Bosomkonsi (the virtuoso)
7. Bosomdwerebe (the eccentric)
8. Bosomayensu (the defiant)
9. Bosomsika (the fastidious)
10. Bosomkrete (the chivalrous)
11. Bosomafram (the kind)
12. Bosomafi (the chaste)

We can only guess what Yaa Asantewaa's *ntoro* was, but I like to think she was Bosomayensu (the defiant). As people come of age, the *ntoro* becomes influenced by the *sunsum*, which gains experience and knowledge.

To sum up, the emotional makeup of Akan individuals is believed to be a mixture of a temperament inherited from their mothers and fathers via *mogya* and *ntoro*. At the same time, things that influence you as you grow up are made manifest through the *sunsum*. The soul has two immortal parts: the *ōkra*, which feels and knows, and the *sunsum*, which desires things and guides the body. It also has a mortal part: the *honam*, or body. These are the foundations on which Akan emotions are and were built.

Feelings, Feelings

Sadly, it's challenging to study discrete emotions as understood in the Twi language. I know of no adequate studies that have been done. Thankfully, Fante, one of the most studied and widespread of the Akan languages, is mutually understood by speakers of Twi. That means we can compare and contrast Fante words with Twi words.

The first hurdle is one I'm sure you're getting used to by now: there are no Fante or Twi words that map precisely onto the English word *emotion*. In fact, many languages that are similar to Fante—such as Dagbani and Ewe—don't have any words for grouping feelings together. Fante has *atsinka*, which in Twi is *atenka*. These terms refer to all internal feelings—hunger, thirst, tiredness, the feelings we call emotions, the lot. If you feel it inside your body, it's an *atenka*.

In a recent study, people who spoke only Fante were asked to list as many of these sorts of feelings as they could think of.[10] Most of these have equivalents in Twi. They thought of sixteen that we'd understand as emotions. Most have something to do with the body. It's probably best explained using a table.

Twi	Fante	Part of the Body	Literal Translation	Closest Translation
Anigyee	Anika	Face: Eyes	Eye-agree and relief	Joy or contentment
	Anigye		Eye-get or eye-deliverance	Excitement or happiness
Anibere	Anibre		Eye-red	Determined or jealous
Amiwuo	Aniwu		Eye-die	Shame
	Anyito		Eye-put	Guilty or ashamed
Asomdwoee	Asomdwee	Face: Ears	Ear-cool	Peace or contentment
Anyimguase	Animguasee	Face: General	Face-floored	Ashamed or disgraced (more than losing face)
Ahooyaa	Anowoyaw	Skin	Skin/self-pain	Envy
Anwonwa	Ahobo		Self (in Twi) and skin-drunk (in Fante)	Surprise
Ahobreasee	Ahobrase		Self, skin-under, and skin in submission	Humility
Abufuo	Ebufo	Chest	Chest-grow or to grow weeds out of the chest	Anger
Ayamkeka	Ayemhyehye	Stomach	Stomach-burn	Anxiety
Adwenemuhaw	N/A	Mind	Modern Twi: problem of mind	
N/A	Akomatu	Heart	Heart-fly	Fright
N/A	Yawdzi	Hunger pangs	Pain-eat or feel	Sorrow
N/A	Bre	N/A		Tired or weary
N/A	Basa	N/A		Agitated, ill, or irritable
Yawdzi	Tan	N/A		Hate

Table 1: Emotions in Twi and Fante

Usually, when Akan people are happy, they feel it around the face; specifically, in the eyes. Happiness is described as something that allows people's eyes—as well as their beliefs and opinions—to meet.

It's known as eye-deliverance, or eye-agree. Good feelings appear to be mutual, shared—something for the wider community.

Old proverbs tell us much the same thing. If an early nineteenth-century Akan felt happiness (eye-agree, or eye-deliverance) that's because "love brings a good manner of living to the home" and "what is enjoyable is enjoyed by all."[11] Pleasant *atenkas* are good for society. They are to be shared in order to improve communal bonds. Although this can go too far. One proverb says, "Too much benevolence brings suffering to the generous." Sharing, but not oversharing, is good for the community.

Bad feelings are different. Pain-eat, or eating when feeling pain or sorrow—something I can certainly relate to—is to be kept to yourself. Proverbs tell the Akan, "He who weeps cannot weep beyond the cemetery" and "A senior man does not give way to grief in public." They warn that "weeping is contagious." That the Akan believed weeping to be contagious might explain why linguists cannot trace a word meaning "sadness." Sadness is a long-term feeling, often impossible to disguise. Sorrow is a personal *atenka*, able to be hidden from the wider group.

Eye-red and the self-inflicted skin-pain of envy are also to be avoided. Envy could cause self-pain or turn your eyes red: "If a man suspects that his wife commits adultery, he should go and suspect the fallen tree across the path to his wife's farm," meaning you should keep your jealousy within limits and perhaps try to find the real reason why your wife is shutting you out. You ought to control your negative *atenkas* and keep your bad feelings to yourself. As the old Akan proverbs say, "If you are hungry, it is only you who feel it" and "If a person is unhappy [pain-eat, or eating while feeling sorrow], it is his own fault."

But you also have a responsibility to help people who are feeling bad. The old proverbs suggest that the early nineteenth-century Akan ought to calm angry individuals down: "We mix hot water with cold water." They say to be mindful of unruly family members: "If you have a dog as a member of your family, you are never without tears."

Should the Akan fail in their attempts to control their bad feelings, there was a whole hierarchy of shame that awaited them, starting in the eyes as guilt, or eye-put (perhaps referring to the inability to make eye contact when feeling shameful, in contrast to the focused eye-get of happiness). Maybe this is what proverbs such as "Salt should not praise itself, saying, 'I am sweet'" are getting at—telling us to avoid the arrogance of envy-born pride. Next was eye-die, a death in the eyes, signifying shame. Finally, there is the face-floored sensation of feeling disgraced.

Anxiety and fright—or, rather, stomach-burn and heart-fly, perhaps meaning heart palpitations—may well play a role in keeping members of the community in check. People on the bottom rungs of society experience anxiety, or stomach-burn, when thinking about taking anything considered a luxury for themselves rather than leaving it for those higher in social rank: "The enslaved avoid the best palm nuts."

Let's pause for a moment and look into another bit of modern emotional theory that I hope will be helpful as we continue. I want to talk about something that comes up quite a lot in the history of emotion, and history in general, called embodiment. In psychology, embodiment is the idea that you need more than a just brain to act, think, and feel. Cognition and emotion are not just the firing of neurons in the brain; they're also the way the body interacts with the environment through the senses. Embodiment is "the processing loops that result in intelligent action," as the person who made the idea popular, philosopher Andy Clark, put it.[12] It's not as strange an idea as some seem to think. Western cultures understand red, embarrassed faces, gut-wrenching anxiety, heart-racing excitement or fear, and sorrowful tears. Historically speaking, we've already explored ancient Islamic ideas tying passions to breath and the many ways ancient Hebrews bound bits of the body to their feelings. It's the same with the Akan, though in perhaps a more direct way. Most of the words the Akan use for feelings aren't just *linked* to the body; they also literally *refer* to a part of the body.

Put Out the Fire

One of the curiosities of Akan society, and indeed the societies of many West African cultures, is that it has a strong sense of both community and what some researchers call enemyship.[13] The Akan are and have long been a collectivist society. A collectivist society is one that focuses on the group rather than the individuals in it. Feelings are not exempt from this collectivism. The feelings of the many outweigh the feelings of the few.

In individualist cultures such as that of the United States, the focus is on the self. You live your life based on your own wants and desires, the underlying idea being that if everyone becomes the best person he or she can be, the whole of society will be improved. To collectivist societies such as the Akan, the opposite is the case. To them, the relationships, the interconnectedness, of humans is what forms the self. You're a human born into a network of social interactions that were there long before you were born and will be around long after you die. You are just part of that, for a brief moment. And this network extends beyond people. It includes the land, the political order of your world, and the spirit realm. All this predates you and will outlast you, so you are raised to fit in.[14]

What is certain is that social ties were significant to early nineteenth-century Akan society. Some of the old proverbs highlight their importance. For example, "If you are in need, one hand lies on the belly of another." Some mention how much worse it was for the poor than the rich: "Even the ocean that is deep has an excess of salt; how much more the shallow ditch." There are also proverbs describing the need for mutual care: "The cat says it is for consolation that it rubs itself against man." Loss of kinship is thought of as painful: "He knew me once, but he does not know me anymore; [it] is bitter."

All the while, the community had enemies. Before the colonial powers came along, the Akan would find themselves in skirmishes with their neighbors over land and resources. These were only avoided by the idea that "an army fears an army."

The proverbs tell us that Akan *atenkas* are, and were, tied to people's behavior, especially around one another. Keeping sorrow inside, despite the hunger pangs thrashing through your body, is done for the good of society. Not becoming chest-grow angry or eye-red jealous is good for society. The Akan have an "anger-in" approach to feelings. To show what are generally thought of as negative emotions publicly would be an act of *anyimguase*, or face-floored disgrace. Well, usually, anyway.

There is an exception, and this is where enemyship comes into play. Groups are more likely than individuals to have enemies. That means individualist societies are less likely than collectivist societies to have enemies. Or at least, that's the theory. In early nineteenth-century Akan communities, the one negative emotion you were allowed to express publicly seems to have been hatred for the other. Dislike of certain groups was part of the collective identity, part of who the Akan were. And as is so often the case with enemies, the worst enemies of the early nineteenth-century Akan were the groups most unlike them: people who displayed bad emotions as readily as good ones.

Anger was and is very important to the Akan, but as you'll see, fear and desire were and are less present. There are certainly elements of the wish to fight, though. In Twi particularly, anger is described as a chest "growing weeds." Once again, this is an embodiment. The chest is where anger lives, where it grows as weeds. This language suggests something unwanted, even dangerous, about anger. It suggests that anger is capable of poisoning the garden of a person's soul and stopping everything else from growing. This might be because, as I've said, anger can kick in before the brain—or, rather, the *ōkra*—has had a chance to control it.

In Akan anger, those weeds, it seems, can be nourished through pressure and heat. An Akan's chest might boil or burn, possibly even feeling as if it were splitting open from the pressure of the rage contained within.[15] There certainly isn't any pleasure taken from anger in the Akan culture. Heat and pressure and the sensation of weeds growing from the chest might be taken as some sort of illness or disease—something wrong, even contagious. Proverbs warn that anger "makes a weak man violent" and that "anger is like a stranger; it does not stay in only one person's house." At the same time, proverbs suggest ways to cure anger: "If someone makes me angry, I can beat him up and recover" and "When someone makes me angry and I insult the person, I recover."[16] These are sentiments I am sure most everyone recognizes. If I were Akan, beating up or insulting someone who made me angry might cause my chest to fall into my belly, helping it cool down.[17] The problem, of course, is that while I'd feel better, the person insulted or beaten up might well catch the disease of anger and so continue the cycle.

The process of anger is a bodily one, to be kept under control. It's only when someone's chest is heated or put under enough pressure that the control exerted by the *ōkra* isn't enough—that the fire burns and the weeds begin to grow. The *ōkra* feels rage. The *sunsum* then decides whether to express it.

Societies with a robust collectivist culture can give rise to a powerful dynamic between so-called in-groups and out-groups. You may remember that the engine that drives attraction, or belongingness, to people in your in-group is fueled by a neurochemical called oxytocin. The problem is that anyone who doesn't cause an oxytocin surge in the same way can become the "other" via a process called othering. This is when we project all the worst traits and habits of our own group onto an out-group and stereotyping begins. The other becomes the worst of us. Catholics become Antichrists in the eyes of Protestants. Protestants become Antichrists in the eyes of Catholics. Manchester United fans become Antichrists in the eyes of, well, everyone else. The effect is that, according to Manchester City fans, at least, Man City players rarely deserve to get booked. That violent challenge was just part of the game, surely? At the same time, Manchester City fans feel that Manchester United players need to be given a red card for even slightly rough tackles. This bias exists because we tend to forgive or even ignore someone in our own group doing something that would be deemed a severe offense when done by the other.

Fight from the Inside

To the Asante, the British were very much an out-group. Yaa Asantewaa had weeds bursting from her chest in a hot rage that needed to be spread to her people. It was right for her rage to infect the Asante chiefs as they sat containing their anger and sorrow, possibly because they thought that was the right way to behave. But the queen mother's royal Asenie *mogya*, and her father's *ntoro* Bosomayensu, meant she was defiant. She erupted in such a way that *ōkras* were stirred and the rebellion could begin.

The Asante began by attacking the British soldiers who were looking for the stool, pushing them back into a stockade. Twelve thousand warriors, weeds growing from their chests, kept those British soldiers, along with a few mixed-race administrators and five hundred Hausa soldiers from Nigeria, trapped in the stockade for three months. The siege continued, but Hodgson and the British somehow managed to escape. Soon more troops arrived from the British Empire. Unfortunately for the Asante, the British army they had fought and defeated in the past had been upgraded and enhanced during the Industrial Revolution. They were a force to be reckoned with. The end was inevitable.

Yaa Asantewaa, queen mother of the Edweso tribe of the Asante

The Asante, despite their rage, were defeated. Yaa Asantewaa was exiled to Seychelles to join her son, and the British annexed the whole region, known then as the Gold Coast. She was lucky. Most of those who followed her into battle were massacred by the reinforcements. But the war had been more uncomfortable than the British liked. They allowed the Asante to pretty much govern themselves autonomously. While not a fully self-governing colony, they were almost entirely left alone, and their customs and system of law were left virtually untouched, at least in comparison to some other British colonies at the time. Though that certainly doesn't mean the British didn't interfere with Asante rule at all.

In 1926 Yaa Asantewaa's son—Prempeh I—returned to his people.

He wasn't allowed to call himself *asantehene*, or king, but beyond that, he was allowed to govern relatively freely. The Golden Stool was so well hidden that it wasn't found until 1921. Sadly, the African road workers who found it got into more than a little trouble after removing some of the gold. They were sentenced to death by the local courts for defiling the sacred object. Luckily for them, the British saw this as an opportunity to interfere and commuted the sentence to banishment. The British, perhaps having learned their lesson twenty years previously, promised not to touch the stool.

In 1931, the Golden Stool once again became the seat of Asante power. That's the year that Prempeh II, the grandson of Yaa Asantewaa and nephew of Prempeh I, sat on it to be crowned the new *asantehene*. You can now find the stool in the Asante royal palace, in Kumasi, Ghana. It's here where the current *asantehene*, Otumfuo Nana Osei Tutu II, still rules with quite a lot of political freedom. It is a freedom that can be traced back to Yaa Asantewaa and her rebellion.

The influence of Yaa Asantewaa's rage extended far beyond the region where she lived and fought. Throughout the world, her name is synonymous with female empowerment, African empowerment, Black empowerment, and bravery in the face of oppression. In Ghana, the nickname Yaa Asantewaa is often given to women who break glass ceilings and overturn assumptions about male-dominated professions. Her memory influences members of the global African diaspora, particularly African Americans who wish to channel her spirit into their lives.

She also still has the power to stir up political tensions in the region. For example, the current *asantehene* boasts a lineage that goes back to the first Asante king, Osei Tutu, the man whose lap first received the Golden Stool. Yaa Asantewaa's descendants comprise a rival lineage. In July of 2004, the Yaa Asantewaa Museum, in the Atwima Mponua District, burned to the ground in what appeared to be an act of arson. Many of the artifacts that once belonged to the late queen mother— including those in the image above—were destroyed. Many people across the political spectrum were quick to blame one another. Well over a century later, these historical tensions remain.[18] The strong *sunsum* that gave *edwesohemaa* Yaa Asantewaa presence, coupled with the weeds that erupted from her chest when her *ōkra* became inflamed with *abufuo*, are still shaping history in West Africa and beyond.

Eleven

Shell Shocks

It so happened not many of our number wore bandages: we bore few signs, outward and visible, that we had been wounded. We were not the battle-stained heroes who had been expected. There was a silence which could be felt. We hung our heads in inexplicable shame. "Let's get off home," a buxomy, loud-voiced dame counselled. "Them's only some of the barmy ones."[1]

This passage, written by a soldier named W. D. Esplin at the beginning of World War I, encapsulates the common attitudes toward mental illness at the time. Esplin was writing about returning home from the war, too emotionally exhausted to continue. Overcome by the conflict, he thought coming home and going to Netley Hospital would bring him relief. The response of the unwelcoming crowd shattered all that.

The problem was that these men had become hysterical, and men weren't supposed to do that. The prevailing theories of the time blamed everything from imbalanced humors to wombs that could wander around the body causing chaos in what was wrongly thought to be a "women's illness." But ever since the onset of the Great War, truckloads of soldiers had been shipped back from the front lines exhibiting a fairly consistent set of afflictions: panic, nausea, blindness, hallucinations, the reliving of stressful situations, and a range of other complaints. Both the Allies (Britain, France, Russia, Italy, Romania, Japan, and the United States) and the Central Powers (the German Empire, Austria-Hungary, Bulgaria, and the Ottoman Empire) had to face what came to be known as shell shock—a mental illness with a wide cluster of symptoms. A 1918 book about shell shock described it as

> a loss of memory, insomnia, terrifying dreams, pains, emotional instability, diminution of self-confidence and self-control, attacks of unconsciousness or of changed consciousness some-times accompanied by convulsive movements resembling those characteristic of epileptic fits, incapacity to understand any but the simplest matters, obsessive thoughts, usually of the gloomi-est and most painful kind, even in some cases hallucinations and incipient delusions.[2]

In addition, there were reported bouts of anxiety, sudden blindness and panic, sleepwalking, and a range of other physical and mental ailments that could suddenly affect fighting men. It was not the same as modern PTSD—its symptoms were much broader, and the underlying causes were thought to be quite different. Shell shock is closer to what modern psychiatry often calls a combat stress reaction, or CSR.[3] Shell shock was immediate, resulting from sudden trauma or shock, whereas PTSD usually occurs months or years after the events causing the trauma. Furthermore, absent from shell shock were the feelings of guilt and aggression often associated with PTSD. Instead, it presented as "somatic discharge"—the discharge of extreme emotions through crying, shivering, and shaking. CSR can lead to PTSD, but the two are understood and treated differently. The long-term effects of shell shock were lumped in with the initial trauma under a single diagnosis. The final difference is the one we'll mostly be exploring below: the cause of PTSD—and CSR, for that matter—is never said to be a lack of willpower, but it was for shell shock. Some people even described shell shock victims as cowards.[4] Shell shock was more similar to a much older emotional illness—*hysteria*.

Attempts to understand this new version of a very old emotional disorder shaped history perhaps as profoundly as the violence of the war that caused it. They spawned the growth of an ever more prominent new field of science: psychology. The application of psychology to the problem of shell shock inspired a reevaluation of what mental illnesses are. The prevalence of shell shock also instilled a need to understand emotions better, for no more significant a reason than that both sides wanted to win the war. Shell shock had to be cured so the men could be put back into the conflict. It changed wartime, and peacetime, forever. But attempts to understand emotions scientifically didn't begin with the first shots fired in 1914.

You'll Be a Man, My Son!

Before we explore the science of emotion as understood at the time of the First World War, we need to understand a something about that era's notion of manliness. For modern Western men, awareness of toxic masculinity and newly flexible beliefs about gender roles notwithstanding, demands to "man up" or act "like a real man" evoke a toughness, a hardness, and the wish to acquit ourselves in the court of manhood. This sort of tough manliness is a surprisingly recent invention. To be sure, ideas about manliness and masculinity have always existed in some form or another. But like any gendered identity, what it means to be a man has been known to fluctuate and change over time.

At the time of Plato, being an active giver, rather than a receiver, was manly. An adult Athenian man's role was to be *active*, to do things like fight in wars and get involved in politics. Women, the enslaved, and boys were expected to be *passive*. They received things given to them by older Greek men. In women's case, this included money to run the household, possessions, gifts, and the like. They weren't supposed to get involved in politics, though in practice, they sometimes did. Meanwhile, pederastic sexual relationships in which the boy, for lack of a better term, was the receiver were normal. Sex between adult males was taboo if for no greater reason than that it renders the question of who is active and who is passive more complicated. A passive, "receiving" adult male might be seen as weak.

The rise of Christianity and Islam borrowed from the ancient Hebrews by associating masculinity with being a breadwinner, a patriarch, the head of one's household. This was a masculinity of class and power, of who held the purse strings and made decisions. A boy king could claim greater manliness than any aged laborer. The powerful could display their passions in ways that poor people couldn't. A handyman had to remain free from extremes of emotion, while a knight might openly weep for a fallen comrade. The Industrial Revolution changed all that. A new sort of working-class man left his home to begin his job as a laborer. Work became dirtier, harder, and more regimented than ever before. It required toughness, discipline, and the willpower to control your emotions at all times.

One of the best descriptions of this new manliness comes from the poem "If—," by Rudyard Kipling. You'll find there's rarely a discussion of late Victorian masculinity that doesn't dip its toe into this work, and there's a good reason for that.[5] The whole poem, written in roughly 1895, is a list of things that make a man a man. It contains such nuggets as:

> *If you can keep your head when all about you*
> *Are losing theirs and blaming it on you,*
>
> ...

If you can force your heart and nerve and sinew
To serve your turn long after they are gone,

. . .

If neither foes nor loving friends can hurt you,
If all men count with you, but none too much;

. . .

Yours is the Earth and everything that's in it,
And — which is more — you'll be a Man, my son![6]

Kipling's guide to manliness applied just as well to almost any civilized European culture at the time. The rise of the machine had brought with it a new ideal, one according to which being a tough, emotionless man was thought to be the epitome of masculinity. What's more, Darwin had come up with a theory that appeared to legitimize Kipling's view, if you looked at it from a certain angle. I'll get there in a moment. But first, I need to make a brief detour to a hospital in France.

Le Hystérie Masculine

At ten o'clock on a cold October morning in 1885, a tall almost sixty-year-old man walked into the lecture theater at Saltpêtrière hospital, in Paris. His eyes were dark, softened by the passage of time. Long wisps of hair fell behind each ear, and his lips were full, protruding above his closely shaved chin. He had the air of "a worldly priest from whom one expects a ready wit and an appreciation of good living."[7] That is, that's how he looked to one of the students attending that day — a young man by the name of Sigmund Freud. Freud knew of the twenty-one case studies that his teacher, Jean-Martin Charcot, had written about hysterical men. These were just the tip of the iceberg. He'd write and publish forty more over the following three years. Before Charcot, no men were admitted to Saltpêtrière. But the growing number of workingmen describing symptoms that seemed—to Charcot, at least—almost identical to the symptoms of women made him curious. Freud would have been watching in excited anticipation. He admired Charcot so much that he had to dose himself with cocaine in order to drum up the courage to talk to him at parties.

Charcot's examination, performed in front of his students, began with two subjects: one man and one woman. He began with hypnosis. First, the patients became relaxed. Their anxiety lifted, and their panic seemed to dissipate. Charcot called this state lethargy. Next, he drew them into a state of deep sleep, or catalepsy. Finally, he put them into a dream state—somnambulism—and began to probe their minds to find out what was wrong. It was this probing that made Charcot famous: his belief that hysteria-like illnesses could be explored in the mind rather than the body was quite a bold step.

With his patients under hypnosis, Charcot would show the onlookers that the man and woman did not have epilepsy but *grande hystérie*. Previously, it had been hard to distinguish one from the other. But Charcot had a trick. Using hypnosis, he would swap the two patients' symptoms. The man would become hysterical, and the woman would behave as if she had epilepsy. This, he believed, proved that the illness was caused by emotional trauma in the mind rather than bodily harm. That doesn't mean that he thought the diseases were caused by the brain exclusively. There was, he thought, a difference between how the illness affected men and how it affected women.

First, he believed that in women, hysteria could be natural—something just beneath the surface ready to break out. But in men, there had to be some sort of trauma that triggered it. Nineteenth-century men working in factories were frequently injured by machinery. One of Charcot's male patients nearly drowned while fishing; another was almost struck by lightning while working the fields; several had suffered some other labor-related calamity. Railway accidents were common.[8] In Britain, these railway-associated emotional breakdowns even had a name: railway spine.

But trauma alone was not enough. The other influence was, Charcot believed, hereditary. Getting a diagnosis was a simple affair if the patient's mother had suffered from hysteria. In those cases, patients had just caught their mothers' affliction. But if the mother hadn't suffered from hysteria, Charcot would look to the father. Maybe he was a drunkard. Perhaps he was a criminal. Or insane. If it wasn't the father, then maybe it was the grandfather or great-grandfather.

As I wrote above, Charcot didn't think that the fact that hysteria was a disease of the brain meant it wasn't caused by a physical malady. Charcot didn't believe in anything as silly as a soul. He was a man of science. No, hysteria had a physical cause, and in this case the cause was a *tare nerveuse*, or a flaw in the nervous system. It could be a chemical imbalance—we still talk of imbalances even now. It could be an intracranial tumor, or it could be some sort of spinal abrasion. The last of these triggers Charcot called "functional dynamic lesions," which, to be frank, meant "lesions we can't find when we look for them."[9] These wounds would be inherited and sit dormant, ready to cause some sort of deficiency, be it alcoholism, criminal behavior, insanity, or hysteria. It might even skip a generation. That's why he was so interested in people's grandparents.

To sum up, Charcot believed men could get hysteria, but it wasn't like women's hysteria. It was remarkable, physical, and caused by the sorts of manly things men did, such as working in heavy industry, fishing, and going out in thunderstorms. Women's hysteria was just women being women. This misogynistic view is not too surprising if you also consider that, with the exception of nursing, the medical profession was almost entirely male at the time. And these weren't just any men: they were manly men, seeking out the dominance of will that made men manly. Some people suggested that hysteria in men meant they were effeminate, possibly homosexual, but Charcot would have none of that. No, it was men—specifically, men who were unfortunate enough to have invisible lesions on their nervous systems—doing men's work that was to blame.

Charcot's pupil Freud had his own ideas about the emotions that might cause shell shock. Freud grouped all kinds of feelings—desires, impulses, moods, even unusual experiences such as déjà vu—under the term *affects*. Affects, he believed, were physical manifestations, something that happened in the body and the brain. They happen when we come across an object. An object, in this case, isn't just a thing, although it can be. But it can also be an experience such as winning a trophy or an event such as walking or going to war. Often, the affect we feel for an object changes or attaches itself to another object. For example, you enjoy a long walk until you're tired, then you enjoy watching a sunset as you rest. You remember the pleasure of the tired-calm link, and that makes you want to do it again.

Usually, the affects we display when encountering an object are what people expect, but just occasionally, they're a bit odd. Often, the oddness is harmless. It might mean that the person reacting to the object is from another culture, or tired, or hungry. Sometimes, it's deep-seated and problematic, and when that's the case, you have to investigate that oddness to discover its origins.

Another vital thing to consider when it comes to the relationship between objects and affects is that we don't attach one to the other at random. Usually, Freud thought, we are either trying to get away from or trying to obtain something from childhood. That's where the idea of the Oedipus complex—a man wanting to have sex with his mother and kill his father—comes from. It's a theory that gives us a glimpse into the emotional dynamics of a mid-nineteenth-century German household, a household that usually comprised a fiercely disciplinarian dad whom a child would relate to through fear, possibly hate, and a caring mom whom a child would think of with love.

Freud firmly disagreed with James and Lange. By the time he had written his 1915 essay "The Unconscious," he, somewhat ironically, didn't think that affects were unconscious. He wrote, "It is surely the essence of an emotion that we should be aware of it." The reasons you might feel an affect, or the idea that causes the feelings you experience, might well be unconscious. But the affects themselves, he thought, are "processes of discharge" that you are aware of at some level.[10]

The most dangerous of these discharges of excitement was anxiety. Anxiety, Freud believed at the time of the First World War, was an outlet for pent-up feelings. It was a gut-gnawing hodgepodge of the bits of emotion you hadn't managed to discharge fully. It's here, in this anxiety, that some mental disorders and, with them, strange reactions to objects can form. Shell shock was one of these peculiar reactions.

Mass Hysteria

Despite Charcot's and Freud's undoubted influence, the most prominent ideas about emotion at the time of World War I were still a little, let's say, old-fashioned. Not everyone had accepted the new concept of emotion as described by Brown, James, and Lange. The problem is that old ideas tend to take a long time to die, and by 1917, they still infected much of the thinking on the subject of feelings.

These old ideas were viewed through the prism of Darwinian evolution as something that looked a bit more like science than religion. Gone was the hierarchy of the soul, and in its place was the pecking order of evolution—a ladder of life. Even this wasn't entirely new. Aristotle's Great Chain of Being ranked all matter in the universe, with rocks at the bottom, God at the top, and everything else in between. To Christians, this chain placed humans—or, rather, men— just one link down from the angels and two from God himself. Women, unfortunately, tended to occupy a link in the chain below

203

that of men. The female of the species was thought, even in the days of the ancient Greeks, to be less evolved. For the record, I personally suspect the opposite is true.

The new Darwinian chain of emotional being also borrowed elements from Aristotle's three-part soul. The most significant difference was that, according to Darwin, humans and animals weren't separated by reason. He argued that the sort of animal intelligence described in this period as instinct was inseparable from intelligence. This meant that every human, no matter how smart, had the instincts of an animal inside him. What made humans special was *volition*. The ability to choose and exercise free will was the supreme achievement of a tripartite mind. Even the simplest creature could recognize things, even if it couldn't plan how to best use them beyond instinct. *Intellect* sat below *volition* in this chain of being. Finally came *affection*, or *emotion*. But, as I alluded to at the beginning of this section, emotions weren't yet a single psychological category, despite the work people had been doing to try to make them one. Emotions, too, had a three-part hierarchy.

An excellent way to get the lay of bygone intellectual landscapes is by finding out what people were taught in a given era. Psychiatrist Robert Henry Cole's 1913 textbook, *Mental Diseases*, opens one such window, being a popular teaching tool in prewar Britain. According to Cole, *affection* had its own hierarchy. Affections were "attracted by what is pleasurable, and…repelled by what is disagreeable, harmful, or painful." Three being a magic number, there were also three types of affection. The lowest was "feeling,…at the foundation of the struggle for existence," which was "in accordance with the primal laws of self-preservation and reproduction."[11] These were the raw sensations: a stimulus followed by a response without a second's thought. They changed the body, increasing or decreasing the dilation of the small blood vessels and therefore affecting heart rate, breathing, and muscle control. In the "insane," Cole suggested, this category of affection could be "perverted so that a painful state is aroused by what should be pleasurable."[12] Only then do you get to emotions.

According to Cole, emotions were not about suddenly realizing that you're blushing or anything like that. Emotions were "connected with the higher mental processes of perception and ideation."[13] These are feelings built on ideas that can be short (*passion*), long (*mood*), or so long that they become part of a person's personality (*temperament*). Curiously, Cole puts temperaments into four humoral classifications—sanguine, phlegmatic, choleric, and melancholic. That's somewhat surprising, since he almost certainly didn't believe in the old theory of humors. Sometimes bad ideas can linger on for a long time.

Emotions are the things to which Cole gave names such as fear, terror, anger, and love.[14] These, unlike feelings, required consciousness—thoughts. He also wrote that the emotions cause madness when they are excessive, deficient, or somehow perverted.[15]

Finally, at the top of the hierarchy, were sentiments. These lofty sensations "differ from the emotions already described in that voluntary reactions involving attention and judgment come into play." In other words, these are the affections that use the will. They include our old friends the moral and aesthetic sentiments as well as the "intellectual sentiments." Intellectual sentiments, according to Cole, "result from the higher emotions brought into action in establishing the *Truth* or *Belief* of any given statement, which is beyond the person's contradiction or doubt." So they're a bit like the moral sentiments, but concern more scientific things. These sentiments are the feelings that justify an action, that make you feel confident you are right. These, too, in excess, deficiency, or perversion, can cause madness. That's how morals are corrupted, justice is subverted, and art can become "grotesque."[16]

Sentiments and the will were what separated civilized humans from animals. And not just animals but also the uncivilized and the insane.

The more reliant a human was on feelings and emotions/intellect at the expense of sentiments/will, the further down a sort of evolutionary chain of being he or she was. Civilized man was at the pinnacle of evolution. To become mentally ill was to slide down the chain, to lose your grip on your self-control, morals, and judgments, and to become uncivilized and savage. As Claye Shaw, the superintendent at Banstead Asylum in Sutton, England, put it in 1904, madness was a "dissolution from the 'highest state of the individual.' "[17]

The striving for control of the will, and the proper use of moral, intellectual, and aesthetic sentiments, was a thinker's way of describing Victorian-era manliness. Women were more in tune with their emotions/intellect than they were with their sentiments/will. That's why, according to the prevailing beliefs of the time, they seemed more susceptible to mental illnesses. The thought that higher rates of hysteria among women might have been caused by oppression never crossed men's minds. And it wouldn't for a long time. But, as we know, the lingering idea that hysteria was exclusively a women's illness was about to come crashing down around them like mortar fire.

This brings us back to the First World War, when thousands upon thousands of men began to exhibit symptoms that to all intents and purposes looked like hysteria. The horrors of war were believed to have diminished their manliness and reduced them to womanlike emotional creatures, "barmy" because of their separation from their sentiments and the shattering of their wills. Or, if you were inclined to follow Freud, these men had all experienced something in their youth that caused anxiety to be discharged. In any case, the world had to accept that both men and women could suffer from this sort of mental illness, regardless of what caused it. It didn't matter how famous you were.

Poetic Justice

In July of 1917, the Speaker of the House of Commons prepared to read out a letter in Parliament. It was a short letter and quickly came to the point. The author stated: "I have seen and endured the sufferings of the troops, and I can no longer be a party to prolong these sufferings for ends which I believe to be evil and unjust."[18] The author claimed that the war had deteriorated from a noble cause to one of conquest and empire, and he wanted nothing more to do with it. The people in attendance that day must have wondered why this offending upstart hadn't just been arrested, court-martialed, and shot for desertion. They must have been frustrated that this message had the patronage of two of the most respected British thinkers of the time: philosopher Bertrand Russell and author John Middleton Murry. These two men were both still highly regarded, and in fact they still are. It was their involvement that propelled this dispatch from the drawer of a commanding officer to the bosom of the press. The powers that be had to do something about the dissatisfied soldier who penned it, because morale is a fragile thing in the best of times. If it had been anyone else, they might have had him shot anyway. Made an example of him. But the author of this letter wasn't just any soldier. It was Mad Jack. A man better known as Siegfried Sassoon.

Siegfried Sassoon was the son of a wealthy Jewish father and a Catholic mother. He was well educated, tall, handsome, and, although he married a woman, gay. Sassoon joined the army voluntarily out of patriotism before the First World War had even started. Just the threat of it was enough to get him into uniform, a uniform he filled with all the manliness one could muster. The name Mad Jack reflected his bravery—sometimes Sassoon was a little too brave. On one occasion, and to the fury of his commanding officer, he captured a German trench by himself using just a handful of grenades. On many other occasions, Mad Jack left his trench on late-night bombing patrols, displaying the sort of manic courage that inspired the men under him and gave confidence to those above him. He even earned himself a medal—the Military Cross—"for conspicuous gallantry during a raid on the enemy trenches." At the time he wrote his letter, a recommendation for the most prestigious military honor the United Kingdom can bestow, the Victoria Cross, sat on the desk of his superiors. Siegfried Sassoon was the poster boy for manly men.

That is, until his dear friend David Cuthbert Thomas was killed. Sassoon was inconsolable, finding it all but impossible to let go of the grief. Thomas's death was the last straw. The blood, the carnage, the horror that had been getting to him for years burst forth. His Mad Jack persona was merely a ruse to cover the deep fear and depression that consumed him. On one trip back to England, he threw his Military Cross into the River Mersey—an act of catharsis to tame one of his darker moods. Then, while resting in the hospital after contracting measles, he wrote that letter. He titled it "A Soldier's Declaration."

The mood in Parliament upon hearing the letter was almost certainly one of fury. Who did Mad Jack Sassoon think he was? What had happened to his manliness? What were the lawmakers to do? The solution was simple: if Sassoon was no longer acting like a man, perhaps he, like so many others, had gone "barmy." So they put him on a boat back to England to be greeted by disappointed ladies, then on a bus to Craiglockhart War Hospital, near Edinburgh.

Sassoon isn't famous now for throwing medals into rivers or even for protesting against a war he had seemed to support so thoroughly. Sassoon is known for his poems. His work was written as a means to release the violent emotions he felt during the war and keep Mad Jack separate from Siegfried. His need to write didn't desert him while he was in Craiglockhart, but it did shift its focus a little. Rather than the physical horrors of war, he gave us an insight into the minds of the shell-shocked.

SURVIVORS

No doubt they'll soon get well; the shock and strain
Have caused their stammering, disconnected talk.
Of course they're "longing to go out again,"—
These boys with old, scarred faces, learning to walk,
They'll soon forget their haunted nights; their cowed
Subjection to the ghosts of friends who died,—
Their dreams that drip with murder; and they'll be proud
Of glorious war that shatter'd all their pride...
Men who went out to battle, grim and glad;
Children, with eyes that hate you, broken and mad.[19]

What's interesting is that Sassoon focused on his fellow shell-shock patients' wish to get better, their wish to get back to the front line. That is, of course, what the government wanted, too. They wanted these men to get better and fight again. Even Sassoon, feeling mentally refreshed, rejoined the army in 1918. According to Sassoon, even those who were broken by war wanted to get back out and win it. But what's also interesting for our current discussion is the last line: "Children, with eyes that hate you, broken and mad." Sassoon doesn't suggest that these men are effeminate or suffering from a women's illness. He says they have become children.

And here is the big twist: one of the odd things about shell shock is how rarely women are mentioned. Women never got shell shock, not even the nurses serving on the front line. They suffered from hysteria. Shell shock was a man's disease, and it was always a man's disease. This distinction allowed shell-shocked men to be thought of not as men with a woman's sickness but as something else. And that brings me back to that famous final line of "If—," by Rudyard Kipling:

And—which is more—you'll be a Man, my son!

Just as Sassoon likened his fellow patients to children, Kipling described what makes a *boy* a man. Kipling's list of the traits of manliness was an inventory of the things that differentiated adult males from children. That is the key to understanding how shell shock was viewed at the time: it wasn't caused by emasculation; it was the result of something Freud would call regression.

Regression was the idea that great trauma could cause you to regress to an earlier stage of development. It was an extension of the theory of recapitulation. This theory is best known for the phrase "Ontogeny recapitulates phylogeny." Put simply, many scientists thought that as the fetus develops in the womb, it changes from a single-cell organism into a protomammal, then transforms into a simple primate. Next, it becomes a complex primate. Finally, it turns into a human. Of course, we now know this isn't remotely true, but it was widely believed. Similarly, Freud described regression as a sort of "involution," or "a return to earlier phases of sexual life."[20]

At the time, it was thought that the pinnacle of evolution wasn't simply the human: it was the sexually mature, civilized adult male. Young boys were still advancing to that stage, and it was only when they reached it that the will and sentiments could be controlled appropriately. Only then could you be a man (my son). But psychologists also believed that not every boy traveled down the road to manhood at the same pace. For some, manhood might come early, but others might not have reached full maturity by the time they were sent off to war. Shell shock, it was thought, had shown the world that the horrors of modern industrial warfare could weed out those men who had not yet passed beyond the emotional maturity of boys, childlike in their lack of willpower and emotional control.

Sometimes, these immature males might appear to be fully grown men until they were shocked back into a childlike state by the ravages of battle. This idea was so popular that it passed beyond the world of science and into popular culture. For example, Rebecca West's 1918 novel, *The Return of the Soldier*, tells the tale of Captain Chris Baldry, a soldier who returns from the trenches shell-shocked and regresses to a state of adolescence, fifteen years of his life wiped from his memory.[21]

From the Ashes, a Phoenix

Wars are terrible things, and in the First World War, the role that mental illness played was significant. The role that the emotions underlying shell shock, and the understanding of those emotions at the time, played in shaping the outcome of the war can never be calculated. But it almost certainly wasn't small. Regardless, the history of psychology as a serious scientific discipline was irrevocably altered by the emergence of shell shock and the desire to understand and treat it. When the era's prevailing understanding of emotion clashed with the reality of shell shock, it provided the first real test of the idea that emotions are products of the brain. It wasn't a test that the leading intellectuals passed, particularly— many of their ideas turned out to be wrong. But then, psychiatry and psychology were young. The good news is that we have gotten better.

Shell shock's modern cousin, PTSD, is taken more seriously with regard to both treatment and prevention. And although our understanding of it is far from perfect, it is unlikely we'd be as far along as we are without the work of those early pioneers of the science of emotion. Any reader who has sought the help of a counselor owes something to these thinkers, scrambling in the dark to keep people fighting. They were almost completely wrong about everything, but they conferred a certain legitimacy on psychology, which allowed it to start getting things right. Well, some things, anyway. It still struggles with emotions, but we'll get to that.

First, we must turn our attention to a different understanding of emotions, one that was emerging on the other side of the globe and was about to wake a sleeping dragon. It would shake its people into revolution and set a nation on a path to becoming one of the dominant forces in the modern world.

Twelve

The Dragon's Humiliation

At the height of the Second Sino-Japanese War, a conflict that coincided with World War II, members of the Chinese Communist Party were busy putting on a series of shows in villages across China. They weren't shows in the singing and dancing sense — though there was a lot of that, too. These were more real. More sinister. Sometime in the early 1940s (no one is sure exactly when), in the village of Zhangzhuangcun, in the southeastern province of Shanxi, one such show was about to begin.

The first act involved calling all local Communist Party members to a meeting. The purpose of this meeting was to organize. Each party member was given a role, a target, and a series of accusations to level against that target. Usually, party members took aim at landowners, officials who abused their powers, and anyone else who used his or her status to oppress. The party members were also given ranks — chairman, first accuser, second accuser, guard, accountant, and so on.

The purpose of the second act was to get "butts in seats" — to mobilize the locals and pull in an audience. The show took place in what was known as a struggle meeting. The idea was that, through a series of well-rehearsed moves, the Communists could get their audience to join in and speak out about the transgressions of the accused. But there was strength in numbers: the larger the crowd, the better the response.

They needed as many of the locals as they could get—if they only had a dozen or so, the theatrical performance wouldn't be successful.

Once the meeting hall was filled, the third act began. Party members led out the accused. The first accusers would lash out. They would scream; they would weep; they would point their fingers in the accused people's faces. The second accusers would do the same, all supported by the chanting of Communist slogans from strategically placed members in the audience. The participants in this show weren't faking these feelings. They were genuine. They had to be in order to have the desired effect. It was a court of raw emotion, where passions were set free.[1]

On this particular occasion in Zhangzhuangcun, however, the crowd remained silent—not one word, not one accusation. They'd all been taught that they should respect the accused. *Li*, or rules of behavior—a notion you may recall from chapter 9—was critical to their culture. The man on trial was the local official, their ruler, their better. Tradition dictated a proper way to behave in his presence. The Communist chairman asked, "Come, now: Who has evidence against this man?" But there was nothing. Finally, the vice chairman lost his temper and slapped the accused man across his jaw, demanding, "Tell the meeting how much you stole." The slap did the trick.

That night, few in the village slept. They had never seen a peasant strike an official before. To them, it was surreal, like watching a play. But they knew it wasn't fiction. The vice chairman's slap had loosened their fears and their tongues. The peasants broke free from the restraints of their emotional regime and joined in.

First, the women confronted their abuser. They often had more to be angry about than the men, because, as so often is the case, they were more oppressed. The men then followed suit with grievances of their own. The accused often fought back. Indeed, accused people who might do so were given preference in the selection process. A whimpering mess of a man might evoke pity. A robust and defiant opponent would only cause more rage. It was all part of the show, an authentic performance designed to elicit an emotional response and electrify the masses. And it worked. If the accused was lucky, the meeting would end with his being led through the streets wearing a dunce cap. If he was unlucky...well, I'm sure you can guess.

This sort of emotional theater was part of a long, hard slog to victory. The Chinese Communist Party (CCP) had existed formally in China since 1921. It had spent most of its time building a network of secret bases in the countryside and rallying people to its cause. Often, members had to take what became known as *Chángzhēng*, or the Long Expedition—sometimes called the Long March. This was a series of retreats to the north and west of China to escape the clutches of the Kuomintang, or Chinese Nationalist Party. Thanks to American assistance, the Kuomintang's forces numbered a staggering 4.3 million men, but they were nonetheless defeated when the better-organized Communists at the Battle of Liaoshi made their overthrow all but inevitable. On October 1, 1949, the People's Republic of China came into being, with Mao Zedong as its leader.

The success of communism in midcentury China, after two thousand years of imperial and then nationalist rule, isn't surprising if you consider what preceded it. China had spent more than a hundred years being downtrodden, invaded, humiliated, and laughed at. Japan and the West had been particularly relentless bullies. Emotions ran high. And that's why struggle meetings were so important. The Communists succeeded in winning the people of China over because, in large part, they understood the power of feelings—and how to manipulate them.

As in Japan, the prevailing understanding of emotions in prerevolutionary China came from a mix of traditions and ideas, most of which go back millennia. Mao and his allies in the CCP understood them well—so well that they eventually came to serve as his secret weapon. But before we get to see that weapon, let's take a tour of the emotions it was loaded with. It all started with a man walking through the Chinese mountains more than two thousand years ago.

The Dao of Doing Nothing

On a warm day in the third century BCE, in the mountains of the Chinese province now known as Méngchéng Xiàn, legendary Daoist philosopher Chang-tzǔ was taking a stroll with friends.[2] He stopped to look at a particularly resplendent tree, its long branches heavy with leaves of the most vivid green. As he stood admiring it, a woodcutter passed by, showing no interest. Chang-tzǔ asked the woodcutter why he walked past such a stunning tree. The woodcutter replied, "There's nothing you can use it for."

"This tree," said Chang-tzǔ, "by its timber being good for nothing will get to last out heaven's term for it, wouldn't you say?"

The woodcutter agreed.

The next day, Chang-tzǔ walked back to a friend's house, where he had been staying. The friend, happy to see his mentor, ordered that a goose be slaughtered so the companions could eat a fine dinner together. The friend's son asked which of their two geese should be eaten.

"One of them can cluck; one of them can't. Which shall I kill, please?"

"The one that can't," replied his father.

Later on, after hearing of Chang-tzǔ's adventures, one of his fellow travelers asked:

"Yesterday, because its timber was good for nothing, the tree in the mountains could last out heaven's term for it. Today, because the stuff it's made of is good for nothing, our host's goose is dead. Which side are you going to settle for, sir?"

A grin came across Chang-tzǔ's face. "I should be inclined to settle midway between being good for something and being good for nothing...Now a dragon, now a snake. You transform together with the times."[3]

Part of this is a dig at Confucianism—almost everything Chang-tzŭ wrote was. He liked to point out that the notion of *li* did more harm than good. Things change, he thought. "Honesty's hard corners will get blunted." The point is that everyone is "now a dragon, now a snake." Categorical rules that claim to govern particular circumstances, always and forever, don't work. There is a better way, he thought. And that brings me to another story.

Hui Shih, one of Chang-tzŭ's closest friends, got the news that Chang-tzŭ's wife had passed away. Usually, following Confucian tradition, the correct response to the loss of one's wife was to mourn for three years. Confucius himself is said to have deemed one man, Zai Wo, immoral for asking if people can stop mourning after just one year. To even suggest such a contravention of *li* meant that his *xin* (heart) wasn't right. His *ren* (humaneness) was off-kilter. His *chih* (understanding of right and wrong) was distorted. Most damningly, his *hsiao*, or responsibility to his family, was utterly jettisoned. *Li* told you that three years was the correct mourning period, so three years it was.[4]

You might imagine Hui Shih's surprise when he arrived at his old pal's home only to find him drumming on the kitchenware and singing. Hui Shih asked:

When you have lived with someone...and brought up children, and grown old together, to refuse to bewail her death would be bad enough, but to drum on a pot and sing—could there be anything more shameful?

As always, Chang-tzŭ had a response:

Not so. When she first died, do you suppose that I was able not to feel the loss? I peered back into her beginnings; there was a time before there was a life. Not only was there no life, there was a time before there was a shape. Not only was there no shape, there was a time before there was energy. Mingled together in the amorphous, something altered, and there was the energy; by alteration in the energy there was the shape, by alteration of the shape there was the life. Now once more altered she has gone over to death. This is to be companion with spring and autumn, summer and winter, in the procession of the four seasons. When someone was about to lie down and sleep in the greatest of mansions, I with my sobbing knew no better than to bewail her. The thought came to me that I was being uncomprehending towards destiny, so I stopped.[5]

Even life is changing and transient. Death and life are *wu* and *yu;* not-being and being. They're two necessary sides of the same thing—the *dao,* or the way. There's no good reason why life is better than death, in Chang-tzŭ's opinion. To forget this and go to extremes is "to go counter to the Power (*dé*) [of the *dao*]; and nothing in the world which goes counter to the Power [of the *dao*] can last for long."[6] Don't become the beautiful tree or the nonclucking goose. The only way is to submit to the power of the *dao,* and for that, you need balance. You need calmness.

The trick is to avoid things that can agitate you. Daoism holds that the five colors, the five tastes, and the five tones—what we see, taste, and hear respectively—are much to blame for our overexcitement. If the pursuit of these takes over our lives, we become stressed. The best way to be, according to a Daoist text called *Daodejing,* "is to be 'like an infant that does not yet smile' and to have 'the heart of an idiot.'"[7] To be unmoved, calm, easy.

I bring Daoism up in part because its influence in China represented an obstacle for the Chinese Communist Party. Daoism was of no use to Mao Zedong. He didn't see the world as something to be simply accepted. He didn't think the state of early twentieth-century China called for calmness. It was time to be a dragon—and a snake. To stick your neck out. To choose to be the goose that clucked, not the tree that stood forever doing nothing. Neither the mind of a child nor the heart of an idiot was of use to him. He had to find a way to get Daoists to let go, or he had to let go of them. And that meant relying on a more visceral understanding of feelings.

The Qi to Emotions

To explain the way the Chinese used to understand their feelings, I first need to try to explain the concept of *qi*. *Qi* is a bit like energy, or the life force. It's something created by all things. It flows through them and between them. Think of the Force from *Star Wars*—though not so much the moving-objects-with-your-mind bit—and you'll be on the right track.

Another way to think of *qi* is in contrast to *li*. *Li*, as I've said, refers to the right way to behave, but it basically means "pattern"—patterns of behavior as well as the patterns of stuff in the universe that are formed and re-formed into various arrangements without ever being destroyed. Or, to quote Alan Watts, *li* is "the asymmetrical, nonrepetitive, and unregimented order which we find in the patterns of moving water, the forms of trees and clouds, of frost crystals on the window, or the scattering of pebbles on beach sand."[8] This concept is linked to behavior because to understand the patterns that form something is to understand what that something is and how it might be used. For example, understanding the various patterns created by water can help you harness water's power by building a waterwheel. Similarly, understanding what various patterns of behavior do and sticking to them can help keep life ordered—keep the waterwheels of society turning as they should.

If you think of *qi* as everything that's not *li* but at the same time flows through and connects the patterns of *li*, then you're in the ball-park. That still doesn't exactly pin it down, but it's about as close as I can get. The truth of the matter is that exactly what *qi* means changes from author to author and from period to period. As it flows freely through everything, *qi* should flow freely through the body. Blockages in *qi* in certain organs, according to ancient Chinese medical texts, are the cause of certain emotions.

Emotions all start in the liver, because this is the first place where *qi* gets blocked, causing the other viscera to become unbalanced and start overproducing their respective elements. By elements I mean wood, fire, earth, metal, and water—the five elements of which, the ancient Chinese believed, all things are made. Each major organ is responsible for the production of one of these elements, and balanced production is essential for life and the good flowing of *qi*.

Once the heart is unable to receive *qi*, its fire becomes too great to control and all joy is extinguished, affecting the *shen*, or spirit, making it restless. As an unknown Confucian medic wrote ca. 502 CE:

> If the spirit is at peace, the heart is in harmony; when the heart is
> in harmony, the body is whole; if the spirit becomes aggravated
> the heart wavers, and when the heart wavers the body becomes
> injured; if one seeks to heal the physical body, therefore, one
> needs to regulate the spirit first.[9]

If the heart is unregulated, it can lead to mood disorders. But shame isn't a factor in Chinese medicine, which is odd given how important it is in Confucianism and Eastern thought in general. Nor, for that matter, are depression, anxiety, and stress—or any other emotion that would be easily recognizable to modern psychology. Instead, the rough equivalents of shame, depression, anxiety, and stress are found merged in what is best described as an emotional disorder—*bēi díe*.[10] *Bēi díe* involves elements of shame and embarrassment as well as aspects of depression. Another word for it might be *humiliation*. This is what happens when the anger in the liver ends up blocking *qi* from the heart, causing joy to fade and *bēi díe* to consume one's soul. Eventually, after too much humiliation, maybe anger is all that's left.

A Century of Humiliation

In 1839, Lin Zexu was writing a letter. A career politician and the son of a career politician, he had risen quickly through the ranks, becoming the governor-general of the provinces of Hunan and Hubei in 1837 before rising to the office of imperial commissioner. The issue for Lin Zexu was the West's desire for Eastern luxuries and, specifically, how they paid for them. The Chinese weren't receiving money for the goods they sold to the British: instead, they were paid in opium. And China had had just about enough.

It started out relatively harmlessly: the use of opium as an ingredient in Chinese tea for medicinal purposes went back to the seventh century. But ever since people started to mix it with tobacco in a substance called *madak*, it had taken on a sinister recreational role. China banned *madak* in 1729, but that didn't stop the opium from coming or the people from becoming addicted. Matters were only made worse when the British ramped up their imports of Indian opium to China in 1781. The whole sordid industry, from grower to consumer, had been monopolized by the British East India Company—a company that was, not for the first time, profiting from other people's misery.

Across China, that misery was profound. Widespread opium addiction was causing serious problems, not least because people were behaving erratically—casting off the obligations of *li* and falling from the *dao*. Opium use in China wasn't just reserved for seedy dens, filled with the down-and-out. It had become the drug of choice for Chinese officials and the wealthy. It was a status symbol—a badge of honor. Something had to be done. Given that the punishment for taking opium was death, this break with proper decorum was a particular concern. So Lin Zexu wrote a letter to someone he thought might be able to help: the empress of the British Empire, Queen Victoria.

He began by saying how good he thought the trade was between their two empires, reminding the queen how much Britain had gained from it.

We of the heavenly dynasty nourished and cherished your people from afar, and bestowed upon them redoubled proofs of our urbanity and kindness. It is merely from these circumstances, that your country—deriving immense advantage from its commercial intercourse with us, which has endured now two hundred years—has become the rich and flourishing kingdom that it is said to be!

Then he got to the point:

But, during the commercial intercourse which has existed so long, among the numerous foreign merchants resorting hither, are wheat and tares, good and bad; and of these latter are some, who, by means of introducing opium by stealth, have seduced our Chinese people, and caused every province of the land to overflow with that poison. These then know merely to advantage themselves, they care not about injuring others! This is a principle which heaven's Providence repugnates; and which mankind conjointly look upon with abhorrence!

And he pointed out:

We have heard that in your own country opium is prohibited with the utmost strictness and severity:—this is a strong proof that you know full well how hurtful it is to mankind.

Then, calling on Queen Victoria's better nature, he ended with a plea:

Let your highness immediately, upon the receipt of this communication, inform us promptly of the state of matters, and of the measure you are pursuing utterly to put a stop to the opium evil. Please let your reply be speedy. Do not on any account make excuses or procrastinate. A most important communication.[11]

Better nature or otherwise, Queen Victoria never received the letter. Even though it was also printed in the newspapers, it's unlikely she read it.

When it became clear that no response was forthcoming, the Chinese emperor demanded that Lin Zexu act, and act he did. He arrested Chinese opium dealers and placed foreign factories and warehouses under siege until they gave up their stocks of the drug, all of which he destroyed. But the British merchants wanted compensation. Neither Lin Zexu nor his emperor, however, had any intention of providing it. They hoped this would put an end to the matter—that the British would see the error of their ways and begin trading in things that weren't destroying the very people they were dealing with. They forgot about the stubbornness and arrogance of the British Empire and its total commitment to unfettered moneymaking.

They certainly had no idea that the British would send warships to get the merchants' money back. But of course Britain was a fiscal-military state. The money made from opium sales and other trading paid for its navy, whose primary job was to keep that money flowing. In a foreshadowing of what was to happen when the Americans reached Japan, the British arrived with weapons that far outstripped anything the Chinese had on hand.

And so began the first of the Opium Wars. The Chinese were decisively beaten, made to pay for the poison the British forced on them, and forced to cede the island of Hong Kong to the British Empire. (They didn't get it back until 1997.) Worst of all, British outsiders had defended Chinese citizens' right to turn their backs on proper behavior. It wasn't just a clash of cultures; it was also a clash of individualism versus collectivism. Foreigners had rubber-stamped the abandonment of *li* in exchange for addiction. It was the start of what became known as the century of humiliation—a nation dishonored, forced into shame, overwhelmed by *bēi dié*.

Over the course of the ensuing hundred years, China would be humiliated again and again. It would lose the second of the Opium Wars, and its emperor's imperial summer palace would be destroyed in the process. It would be defeated by France, Japan, and Russia, resulting in even more loss of its territory. What's more, internal struggles broke out. An anti-foreign and anti-imperialist rebellion in 1899—the Yihetuan Movement, or Boxer Rebellion, as it was known to the British—was quashed not by Chinese forces but by an invasion of eight nations: the United States, Austria-Hungary, Great Britain, France, Germany, Italy, Japan, and Russia. Then, during the First World War, Japan issued a list of twenty-one demands, the *Taika Nijūikkajō Yōkyū*, giving it even more power over parts of China and its economy. Even though China had managed to become a republic in 1911 through the *Xīnhài Gémìng*, or Xinhai Revolution, it still had no teeth. The powers that eventually curbed Japanese control over China were foreign—specifically, Great Britain and the United States. This was yet another humiliation.

Then World War II came along. It began for China, some argue, in 1931, when the Japanese invaded Manchuria, an area to the northeast of China. When China appealed to the West for help in defending its territory, the response was little more than a slap on Japan's wrist, leading to its withdrawal from the League of Nations. Japan got to stay in Manchuria. Even more humiliating for China, Japan installed the emperor the Chinese had kicked out of power in 1911—Puyi—as the puppet ruler in the province.

Conflict resumed in 1937, when Japan launched a full-scale invasion of China. As Japan's role as an Axis power grew, allied as it was with Italy and Germany, this conflict was sucked into the tornado of the Second World War. Once again, the West helped a bit by providing sufficient resources to keep China fighting to the end. But when the end finally came, it had little to do with China. Japan only withdrew from China when, for the second time in its history, it was brought to its knees by American technology. This time it wasn't ironclad ships, though; it was the atomic bombs that turned the cities of Hiroshima and Nagasaki into apocalyptic wastelands.

The Emotional Revolution

I tell you all this because, without some idea of the shame, the forced dishonor, and the humiliation China suffered over the course of this long period in history, it's difficult to understand how and why the Communist Party's tactics worked.

There's a lot of debate about how the party got the better of the Kuomintang. More often than not, historians argue that it came down to something painfully mundane—such as the Communists outnationalizing the nationalists or having better ideas about land reform. Some cite the way the Communists used symbolism to imprint themselves on the minds of the people. Others think it was a matter of organization. The Communists were simply better organized than the Kuomintang. But for everything these explanations get right, each one misses an important point.

A key aspect of the Communists' strategy was a technique known as *tigao qingxu*, or "emotion raising." Emotion raising involved techniques such as *sixiang gaizao* (thought reform), *suku* (speaking bitterness), *piping-ziwo piping* (self-criticism), and *kongsu* (denunciation).[12] It also involved older techniques—martial arts routines, meditation, oaths, and other ways of creating a new and better *dao* and, with it, new and improved rules of *li*. The trick for the Communists was to show the people that the *li* of the nationalists was as off-kilter as getting over a bereavement too quickly. They needed to unblock the nation's *qi* and cure it of its century-long bout of *bēi diē*.

Mao Zedong certainly knew about emotion raising and the best ways to direct people's feelings toward another path. In addition to the struggle meetings, the CCP used actual song-and-dance shows as part of its quest to win minds and, more important, hearts. The Communists put on plays depicting landowners and officials in ways that evoked strong emotions — laughter, anger, sorrow. One journalist interviewing the Communist soldiers after the Long March discovered that

> there was no more powerful weapon of propaganda in the Communist movement than the Reds' dramatic troupes, and none more subtly manipulated... When the Reds occupied new areas, it was the Red Theater that calmed the fears of the people, gave them rudimentary ideas of the Red program, and dispensed great quantities of revolutionary thoughts, to win the people's confidence.[13]

The plays were a device designed to provoke a thirst for revenge against the nations humiliating China and pushing it off its path. Obviously, the Communists' approach involved more than just plays. As Professor Elizabeth Perry has pointed out, "[The play] was, in a sense, a metaphor for the entire enterprise."[14] Even the land revolutions, when peasants were urged by the Communists to seize any property their landlords possessed during the latter stages of the Second Sino-Japanese War, were a kind of emotional theater, as were the struggle meetings. And the theater didn't stop with the success of the Communist armies.

China Stands Up

On September 21, 1949, just after the Communist Party had become the clear victors over the Kuomintang in the Chinese Civil War, Mao Zedong made a speech. In it, he said, "We are all convinced that our work will go down in the history of mankind, demonstrating that the Chinese people, comprising one quarter of humanity, have now stood up." He promised that "no imperialists will ever again be allowed to invade our land" and declared, "Let the domestic and foreign reactionaries tremble before us! Let them say we are no good at this and no good at that. By our own indomitable efforts we the Chinese people will unswervingly reach our goal."[15]

Mao was acknowledging the humiliations of the previous century. His speech's goal was to lift the Chinese from a mass sense of *bēi díe*, to move them on to something better. Even here, Mao was using the techniques of emotion raising to construct a new emotional regime, new patterns, new *li*. He kicked off a continual Chinese quest for international status, for honor, for respect. But he hadn't yet eradicated the humiliation from his own country.

A short decade later, in 1958, Mao began his Second Five-Year Plan, also known as the Great Leap Forward. Sadly, it was a giant step backward. The purpose of this leap was to modernize, leaving the agrarian agricultural economy behind for something much more industrial. One unfortunate side effect of this was that officials in rural parts of China sent more produce to Beijing than was needed to sustain the locals. That led to the worst famine in human history, one that killed between fifteen and forty-five million people.

Following this catastrophe, Mao changed direction. He wondered if it might be easier to transform China through a Great Proletarian Cultural Revolution, as it came to be known. The plan was to go back to the process of emotion raising that had served him well in the past and shake off the last dregs of humiliation. The Cultural Revolution began in 1966 and was carried out primarily by the nation's youth. Rather than an explosion in art and literature, as you might expect, it was about changing the culture at a basic level—once and for all cementing a new *li*.

The techniques that had been used in the early days of the revolution came back into play, but this time more often in schools and universities. Struggle meetings were replaced with "mass criticism sessions," in which people suspected of capitalist leanings—usually middle-aged men—were put on display and humiliated. According to an eyewitness named Liang Heng, they went something like this—and fair warning: it's not exactly pleasant.

> The loudspeaker called us all outside, and in a few minutes I saw it coming. A group of Rebels were in the lead shouting "Down with the Capitalist Roaders" and "Long Live Chairman Mao Thought." Following them were about ten of the old "leading comrades" tied together on a long rope like beads on a string, their hands bound. They were wearing tall square-topped paper hats inscribed with phrases like I AM A BASTARD or I AM A FOOL, and around their necks were wooden signs with their names and crimes...
>
> The Capitalist Roaders [knelt] on the platform, their hands tied behind their backs with long ropes...The meeting went on and on, and whenever someone stumbled there were cries of "Give him an airplane ride, give him an airplane ride!" At this the Rebels tossed the rope binding the man's arms behind him over a pipe at the top of the auditorium and hoisted him up in the air, letting him squirm in agony like a dragonfly with pinched wings.[16]

The role of cruel mass-criticism sessions was the same as that of struggle meetings: to free *qi* and redirect *li:* to impose a new emotional regime. That Mao had no truck with traditional Chinese medicine would in no way have affected the billion people in China who always believed in *qi*. To them, letting go of anger freed a route back to joy through the release of pent-up *qi*.

These sessions were the last gasp of a century of humiliation—a final reclaiming of China's honor through the transfer of shame onto its enemies. As Elizabeth Perry points out, "The Cultural Revolution showed yet again, if further proof were required, just how volatile and fluid individual emotions can become in the context of group politics."[17] In many ways Chang-tzŭ was right: things do change. Especially political tides. Emotional regimes, patterns, and *li* are not and cannot be set in stone. If there is something universal about emotions, it's that they ebb and flow—a bit like *qi*.

The drive to get past the century of humiliation, both internally and internationally, still motivates China today. It has since become one of the wealthiest and most powerful countries on earth, and that, in no small part, is the result of an emotional tide that first began at the beginning of the century of humiliation. It was inevitable, given its people's understanding of feelings, its vast population, and its immense resources, that China would break free from its downtrodden status to become the economic and military powerhouse it is today. A wounded dragon can only sleep for so long.

Thirteen

Love and the Mother(land)

It's September 12, 1962, at Rice Stadium, in Houston, Texas. On this day Rice University's football team, the Rice Owls, is neither playing nor training. The cheerleaders aren't there, and the crowd is quite a bit older than it usually is. That's because today President John F. Kennedy is giving a speech, and he has promised that he'll be saying something momentous. Kennedy had overcome the lead of his opponent, Richard Nixon, two years earlier. His campaign's whole message had been about rekindling a love for the United States. In his inauguration speech, Kennedy said: "Ask not what your country can do for you—ask what you can do for your country." He was challenging every US citizen to be proud to be American, to love America, and to work together for a golden future. At the same time, he was reinforcing America's hatred of communism and the Soviet Union, continually referring to its citizens as "them." The goal of this messaging was to use a collective hatred of the USSR to bolster a set of intergroup emotions based around love for the United States.

To America's surprise, it had been the Soviets who had first managed to send a vehicle to outer space. The real starter pistol of the Space Race was fired when a satellite called *Sputnik 1* was launched into orbit on October 4, 1957. The USSR had previously been shy about its rocket science—in part because it wanted to shock America and in part because things kept going wrong. When the Soviet government finally revealed what it had done with the Sputnik program, officials were amazed by the reaction in their homeland. Comrades from across the country were excited and proud, and it didn't take the state long to tap into this emotional outpouring by designing propaganda around its extraplanetary achievements. The government asked the people to look out for the satellite as it passed overhead (although it was all but impossible to see) and to tune in to its now famous "beep, beep" signal. Propaganda posters bearing uplifting images and slogans such as "We were born to make the dream come true!" and "Sons of October—pioneers of the universe!" were printed and plastered around major cities. The Soviet Union was suddenly very proud of its space program. Its intergroup bonds were made tighter as national pride swelled.

America was aghast. Ever since the Cold War began, Americans had been told not to worry because the United States was far superior technologically to those "backward Commies." Finding out that this wasn't entirely accurate snuffed some of the fire from the rocket engines of US national pride. Something had to be done to restore it. America had to get its own satellite into space. By 1958 it succeeded, not long after the Soviets had launched their second satellite, *Sputnik 2*, to circle over the heads of the decadent Westerners.

The newly reelected president, Dwight Eisenhower, knew he had to up America's game. So a few months after America's first successful satellite launch, he set up the National Aeronautics and Space Administration, better known as NASA. The organization's task was to restore national pride—or, rather, alleviate fears—by proving America's technological superiority to the USSR. It planned to do this by sending a human into space and then getting him home again without killing him. It was all set up. NASA developed a new rocket at great expense especially for the task—part of Project Mercury—and was about to make Alan Shepard the first man to visit space.

Then the Soviets surprised them again. In April of 1961—one

month before the planned US launch—they quietly and secretly put Senior Lieutenant Yuri Gagarin of the Soviet Air Forces into orbit around the earth and brought him home safely. Before too long, Soviet officials were bragging about their achievements to anyone who would listen. Once more, Americans were stunned; for the second time, they were the runners-up. Americans really don't like being runners-up. US pride was badly damaged, and a concern that the Soviets might also be ahead militarily began to grow. The new US president, John F. Kennedy, had only one option available to him if he wanted to regalvanize love for the nation. He had to shoot for the moon.

When Kennedy took the podium at Rice Stadium in 1962, he would make perhaps the most ambitious scientific promise in history. No matter what it took, the first man on the moon would be an American. He said:

> We choose to go to the moon. We choose to go to the moon in this decade and do the other things, not because they are easy, but because they are hard... That challenge is one that we are willing to accept, one we are unwilling to postpone, and one which we intend to win.[1]

JFK made the race to the moon more than just a quest for technological brilliance. For him, it was a battle between good and evil. He wanted America to place a "banner of freedom and peace" on the moon before the Soviets could plant a "hostile flag of conquest." Of course, those hateful Soviets only wanted to get to the moon so they could occupy and own it. They wanted to expand their Communist empire into space—or at least that's what the average American believed at the time, thanks to JFK. The moon shot was JFK deliberately mixing love for the United States with a hatred of the enemy.

Unfortunately, Kennedy didn't live to see men land on the moon; he was assassinated six years before his dream became a reality.

Nevertheless, at 4:17 p.m. Eastern Daylight Time on July 20, 1969, with the whole world huddled around millions of television sets, the Apollo 11 mission made good on Kennedy's promise. After a journey of 240,000 miles that took just over four days, Americans set foot on the surface of the moon. While there was a sense that the whole world had come together to share in this moment, the moon landing had its biggest impact in the United States. It triggered an outpouring of pride and love for the United States of America. American newspapers and politicians gushed with enthusiasm. The *New York Times* declared it an "ancient dream fulfilled," and President Richard Nixon claimed that "for every American this has to be the proudest day of our lives."[2] Reaching the moon had accomplished, if only briefly, what JFK had set out to achieve. It was not just a technological achievement but also an event that further inspired the desires he was aiming for.

"I Am Become Death"

Why was JFK so fixated on getting to the moon? Well, imagine a beautiful morning in the mid-1950s. The New York skyline is cutting its usual silhouette against the ball of the rising sun, heralding the morning. Sharp, angular black shadows crisscross streets and buildings, rooftops, parks, and squares. This morning, the sun rose more rapidly than usual. Unnaturally so. It was no slow creep over the horizon. This time, the darkness became a blinding flash of intense heat instantaneously. That commanding, timeless skyline began to crumble. This wasn't the sun, and this wasn't the morning. It was the furnace of a nuclear bomb—the atomic wars had begun.

On New York's streets, devastation reigned. As journalist John Lear described it:

> Babies were wailing, women were screaming, and here and there men's voices rose in a babel of confusion. With a roar that

rattled the neighborhood, two huge gas tanks down the river threw up flame like Roman candles. Across the river, on New-town Creek, great piles of lumber were burning. But it was over the rooftops, far down in the direction of Chinatown, that he saw the worst of the disaster.

Great waves of purple and pinkish brown billowed across the city. Hundreds of feet high, they surged up like an angry sea; the powdered ruins of thousands of brownstone tenements. And beneath and beyond the waves glowed the ominous red of fire.[3]

Of course, what Lear was describing, and what I am describing, never happened. New York has never been attacked by a nuclear weapon. This is all from a story written for *Collier's* magazine just one year after the Soviets tested their first atom bomb, in 1949. It wasn't the only attempt to imagine nuclear annihilation at the time or afterward.

I remember the 1984 movie *Threads* vividly.[4] It depicted the reali-ties of a nuclear conflict—albeit with more survivors than was likely. This particular film was all the more terrifying to me because it was set in my childhood hometown of Sheffield, England. Watching my town hall get blown to pieces sent a shiver down my young spine, and it's still shivering. And I am not alone.

Just about everyone who lived during the Cold War was a little bit scared all the time. America's elites knew this only too well. They had to find some way to reduce that fear and stop what they called the "problem of panic." If Americans and their allies were scared about nuclear war, then they might oppose having nuclear weapons. This would not do. It would be as good as handing over the world to the enemy, they thought. Emotions had to be managed. Feelings had to be kept in check. Of course, finding out that the Soviets could put things in space that might drop on you at any second—while the Americans could not—didn't help ease the sense of panic.

The Space Race was a product of two major fronts in the Cold War. The first was about those terrifying nuclear bombs and the need for the two superpowers to show their technological superiority. The second front was emotions. In 1954, American psychologist Howard S. Liddell wrote, "The primitive forces of man's emotions are more dangerous and more devastating than nuclear fission. Who can doubt that the central scientific problem of our time is the problem of emotion?"[5] He wanted to know how to stop people from permanently cowering under their desks in fear of the bomb. It turns out that you do this by controlling people's emotions using a smart little trick called *emotional coupling*.

What Is Hate (and What Does Love Have to Do with It)?

Hate is an interesting emotion: fMRI studies struggle to find a single specific pathway or neurochemical responsible for it. The best candidate is what researchers call the brain's "hate circuit," involving a combination of the insula (thought to be implicated in disgust) and those areas that link aggression to decision making—the putamen, frontal cortex, and premotor cortex.[6] The neuropsychological gap between disliking something because it revolts you and because it makes you want to punch it seems to be narrow. That makes some sense—we do tend to hate people whom we find disgusting. Manipulating revulsion to generate hatred is a technique with terrible power, which the Nazi Party demonstrated all too well.

Hatred also makes us want to either fight or avoid people we despise. If this feels like déjà vu, it's because, once again, we find ourselves at a decision point between fight and flight. Unsurprisingly, our old friend the amygdala also plays a role in hatred.

Hate is a source of "othering"—it was hatred that fueled Yaa Asantewaa's rage at the British. It is linked to things that build our idea of the "other"—stereotyping, scapegoating, and fear of the unknown. It seems likely that hate is a sum of other emotions, a constellation of feelings dependent on context and culture. Yoda was wrong—fear doesn't lead to anger, hate, and the dark side; the feelings that take you to the dark side occur all at once. He also missed one of the most powerful and perhaps unexpected routes to hate: love.

It's often said that there's a thin line between love and hate. You may also recall that Saint Thomas Aquinas thought that love was the opposite of hate. He was probably wrong. Love is not the opposite of hate; it's more of a companion. Long before Aquinas, the ancient Greek poets noticed that love and hate had quite an intimate relationship. It was a mixture of the love of a woman (Helen) and hatred for the man who took her away (Paris) that launched thousands of ships and the Trojan War. Two thousand years later, Sigmund Freud noticed that not only can love and hate be intimately linked but also that some things cause love and hate simultaneously. He thought the reason for this, as usual, came down to breastfeeding and toilet training.

Freud, if you remember, thought all affects stem from things that happen during childhood and a need to either get back to them or away from them. Sometimes, he thought, both impulses occur simultaneously. We might develop a love for breasts as a food source and derive pleasure and relief—if we're honest—from urinating and defecating. But we might also be taught to hate breasts and the feces we produce because they are, respectively and in Freud's view, morally and physically disgusting. He called this mixture ambivalence.[7] While his ideas about the cause of the link between love and hate have been pretty thoroughly debunked, there have been more recent studies of the brain that suggest that he, and the Greeks, weren't entirely wrong.

As it turns out, the part of the brain that lights up when we look at pictures of people we love is the same as the part that lights up when we look at pictures of people we hate. That bit of the brain, the insula, appears to be the bit that decides how strongly we feel about something. It seems to act as a junction that hate and love—and disgust, for that matter—all pass through.[8] This seems like a satisfying explanation.

Usually, when we move from one emotion to its opposite, it's part of a process known as an *emotional chain*, or *segue*. Hope becomes concern; concern becomes worry; worry becomes fear; fear becomes panic.[9] Two opposite emotions can also exist at the same time; what was once called ambivalence is now known as *emotional coupling*. Coupling usually connects two similar emotions—such as the fear and disgust some people feel when they see spiders. But thanks to the insula, it would seem that love and hate can couple, despite being "opposites."

If and when pairs of opposite emotions combine, they form what has been called *moral batteries*.[10] The poles of these batteries can flip because of external forces such as sudden shocks or extreme emotional pressures. Love and hate form a perfect example of a moral battery—a person can flip alarmingly quickly between them when exposed to some sort of trauma or stress. Usually, the change is disastrous, as evidenced by the thousands of crimes of passion on record—not to mention all those Greek ships that set out for Troy.

That one emotion can link to its polar opposite has been known since long before we were stuffing university students into claustrophobic cylinders to look at their brains. You might recall that it fueled Plato's exploration of feelings and the soul, and Shakespeare noticed that "parting is such sweet sorrow" quite some time before the first fMRI was built.[11] Cold War–era American and Soviet leaders knew it, too. They knew that if they wanted nuclear weapons to be justified, even accepted, then the enemy had to be scarier and more hated than the warheads. An excellent way to do that was to find ways to get people to love their countries and hate their enemies. And so a psychological boom began.

Loving America

You may recall a previous encounter with love. A Christian, Augustinian love. A love whose role was to guide you up the stairway to heaven.

You might remember that some modern scientists split love into three types—*lust*, *attraction*, and *attachment*, each triggered by its own cocktail of neurochemicals. But some governments and scientists knew of ways to manipulate love long before anyone understood which chemical did what. America's approach to inspiring love was to emphasize the "American way of life"—freedom, democracy, baseball, apple pie, and all that good stuff the Soviets didn't have. To learn how to do this effectively, the Federal Civil Defense Administration (FCDA) got involved in psychology. More specifically, it got involved in emotion management. This management consisted of setting boundaries: Deciding what people ought to feel. In what circumstances they ought to feel things. What the meanings of those feelings should be. How they ought to be expressed. How best to control them.[12] This was an attempt to create an emotional regime from the ground up.

The Second World War ushered in an era when working together for the good of the American nation was critical. To borrow a construction from JFK, it mattered not what your country felt for you but what you felt for your country. After the creation of the atom bomb, people felt like their country might just kill everyone. The political leaders believed that had to change, and a focus inward, on personal emotions, became essential. Fierce individualism, rather than the collectivism of the war effort, became the best thing for the nation.

In a magazine article published in 1953, the head of the FCDA—Frederick Peterson—put it this way:

> If you are prepared and remain calm you will be performing a service of tremendous value to yourself and your country—and probably the whole free world.[13]

To develop a framework for this new type of large-scale emotion management, enormous amounts of money previously spent on research into politics and economics were diverted into other social sciences.

Companies such as the RAND Corporation were given cash to spend on research into education, childcare, and social welfare, among other things. It's no coincidence that this era saw school counseling introduced, in the 1958 National Defense Education Act. Nor was it mere happenstance that the number of adults seeing therapists rose during this time. The influx of funding helped the American Psychological Association grow its membership dramatically and, in turn, conduct ever more new research. Dozens of new ideas and disciplines sprang up, so many I couldn't possibly list them all here. Professor Dr. Kelly A. Singleton has described this period as the "psychological turn," and I'm inclined to agree with her.[14]

But despite all appearances to the contrary, America's new managers of national emotion weren't interested in changing anything. The primary goal, really, was to take emotional regulation away from the entities that controlled it—namely, religions—and give it to the state.

In 1960 a sociologist named Daniel Bell, who also happened to be one of these emotion managers, wrote:

One might say, in fact, that the most important, latent, function of ideology is to tap emotion. Other than religion (and war and nationalism), there have been few forms of channelizing emotional energy. Religion symbolized, drained away, dispersed emotional energy from the world onto the litany, the liturgy, the sacraments, the edifices, the arts. Ideology fuses these energies and channels them into politics.[15]

That was it in a nutshell: appropriate the existing emotional landscape and redirect its desires toward a political ideology—"the American way of life."

One way to do that was to blur the lines between church and state. Just a little. In 1954, Congress voted to change the Pledge of Allegiance. No longer was America "one nation, indivisible." Instead, it became "one nation, under God, indivisible." In 1956, the national motto also shifted. The previous motto, *E pluribus unum* (Out of many, one), was replaced with "In God we trust." That new motto was added to all paper currency soon after, a constant reminder of Cold War America's divine moral stance in every person's pocket. The implication was that the morality of the United States was somehow endorsed by the Almighty himself. It allowed morals and emotional control to move from the pulpit to the politicians. Emotions were a state-sponsored commodity—part of an all-powerful fiscal-military-emotional state. But commandeering God wasn't enough. Legislators decided to go for an even higher power: moms.

The American Way of Life

Margaret Mead was an anthropologist who was certain that emotions were almost surely created by cultures—I'll talk about her more in the following chapter. In 1942, she released a study called *And Keep Your Powder Dry*, which devoted a great deal of its content to the influence of parents on their children.

> Just as one way of understanding a machine is to understand how it is made, so one way of understanding the typical charac-ter structure of a culture is to follow step by step that way in which it is built into the growing child.[16]

She wasn't wrong. Raising children is known to be the main pro-cess through which a culture, including its expectations about express-ing emotion, spreads. Of course, there's a feedback loop. Your culture influences how you raise your kids, which affects your culture, which . . . and so on. Mead was aware of this, but she thought that mothers are of particular importance when it comes to the advancement of culture because mothers spend much more time with their children than anyone else does.[17] Mead wasn't the only scientist to think this. A great deal of research produced by the RAND Corporation came to the same firm conclusion: American moms know best, but they could do with a little guidance.[18]

One of the issues was that during the Second World War, women had become used to working. They built the bombs and riveted the ships and painted the tanks. Furthermore, the technological developments of the 1950s and 1960s undermined most arguments in favor of women staying home rather than working. It might seem deeply misogynistic now, and it was, but one of the reasons—or, I should say, excuses—for the division of labor in the household had to do with tasks like laundry, cooking, and cleaning. These could take up the entire working day and beyond. Washing machines and other gadgets changed that dramatically. Suddenly there were no real barriers stopping women from pursuing a career. But the government wanted moms to stay home with their children, and so it fought back with compromises.

Books such as Mary and Lawrence Frank's *How to Be a Woman* suggested that women staying away from their children for too long might cause a "real crisis." Their solution? Part-time work, preferably from home.[19] Despite a growing body of magazines, books, and other media dedicated to female empowerment in the 1950s and 1960s, there was just as much emphasis on a different, more regressive, message: "Women, go home and get back in the kitchen. You could even work from there!"

There's an argument to be made that the focus on mothers staying home was a landgrab of labor by men, the patriarchy attempting to reassert itself. But it's hard to ignore its connection to other political aims of the time. America was afraid that its social order was in trouble. The nuclear family was disintegrating as divorces, rates of alcoholism, and the number of people admitted to asylums rose. In the end, the government's solution was to appeal to what it thought was a deep, evolved instinct, a drive shared by all humanity and placed in our brains by the process of biology: *love*.

The powers that be were certain that the biological, natural love a mother felt for her children was the key to ending fear. In a 1951 report for the World Health Organization, British psychologist John Bowlby wrote about the dangers facing a child deprived of motherly love.

Partial deprivation brings in its train acute anxiety, excessive need for love, powerful feelings of revenge, and, arising from these last, guilt and depression. These emotions and drives are too great for the immature means of control and organization available to the young child (immature physiologically as well as psychologically). The consequent disturbance of psychic organization then leads to a variety of responses, often repetitive and cumulative, the end products of which are symptoms of neurosis and instability of character. Complete deprivation...has even more far-reaching effects on character development and may entirely cripple the capacity to make relationships.[20]

His cure for this danger was simple: mothers should not deprive their children of attention. He wrote:

The child needs to feel he is an object of pleasure and pride to his mother; the mother needs to feel an expansion of her own personality in the personality of her child: each needs to feel closely identified with the other...*It is for these reasons that the mother-love which a young child needs is so easily provided within the family* [my emphasis].[21]

The (unintentionally appropriately) named "nuclear family" was essential, which at the time meant a male breadwinner, a homemaking mother, and their 2.4 children. The white picket fence was optional.

There was no better way to reduce the fear of total annihilation than motherly love, legislators thought. Naturally nurtured and loved children would be raised happy and strong, with love for the American way of life and all the free people of the world. They would also righteously and zealously hate people who opposed their way of life, thanks to emotional coupling. To do anything else, according to much of the research conducted during the period of the psychological turn, was to go against evolution. To fight instinct. To subvert human drives. The secondary implication, of course, was that the Soviets didn't love their children, that they fought against nature and the evolutionary status quo to create monsters—not people who loved their way of life but who were frightened of speaking out against it. Cogs in a machine. Unwitting parts of something greater, more sinister. They weren't entirely wrong.

Did Soviets Love Their Children, Too?

It's the last week of June in 1950, and you've been asked to attend a joint session of the Academy of Sciences of the USSR and the USSR Academy of Medical Sciences. It's not like you had much choice: the invitations were sent out by Stalin himself. They were less a request than a chance not to be executed. The reason for the session is to work out where Soviet psychological research is heading. Or at least, that's what its purpose is on paper. In reality, it's to confront scientists who disagree with Ivan Pavlov.

Pavlov, who'd died fourteen years before the meeting took place, was a famous Russian physiologist whose research on the reflex system revolutionized psychology. He showed that people could be conditioned to behave in specific ways. For example, if you start regularly ringing a bell just before feeding your dogs, the dogs will come to associate the sound of the bell with food. Eventually, just hearing the sound of the bell will make them salivate as if they are getting dinner, even if they aren't. This became known as "classical conditioning."

Stalin wanted to use a similar technique on humans. He thought it was a way to produce collective behavior—the sort of thing the Soviet regime liked a lot. But humans aren't dogs, and our responses can be more complicated and unpredictable than theirs. Or at least that's what you would've thought until your colleagues were arrested for "anti-Pavlovian, anti-Marxist, idealistic, reactionary science" damaging to Soviet psychiatry. Once arrested, they were forced to admit they were wrong or be killed.[22] Such actions tend to focus the mind and help you realize that Pavlov was right after all. But this meeting wasn't just about rubber-stamping Pavlov's ideas. It also had a bigger goal—how best to get people to love the Soviet Union.

The psychologists who toed the line faced the problem of finding ways to remove the remnants of a Russia whose thoughts and feelings still leaned toward old, capitalist ways. This wasn't a new issue, although a dislike of psychoanalysis for political reasons had left it a little dormant for a few decades.

In 1924, Russian psychologist Aron Zalkind published a book of his papers—*Essays of a Culture of a Revolutionary Time*. The papers mixed psychoanalysis with radical Communist ideas to create "pathological Marxism."[23] Zalkind argued that physiology and class were linked, and that this linkage needed fixing and evening out. A "new man" needed to be constructed.[24] But how might the government build the new Soviet man,[25] a man who loved the Soviet state even more than he loved his mother—and, more important, a man who hated the capitalists more than anything?

One of the issues was that the Soviets didn't much like Darwinian evolution. They thought that natural selection meant unregulated competition for resources and that unregulated competition for resources in turn allowed the species with the greatest number of unfair advantages to win. This was biological capitalism, laissez-faire nature. It was no good. To the Soviets, communal, socially organized progression—communism—had to be what governed nature.

Similarly, they believed that humans weren't born with inherited emotional states that helped them compete. They were born with a Pavlovian blank slate. This was the Soviet approach to emotion. Emotion was a form of conditioning. They believed children could be classically conditioned to have particular emotions at particular times, leading to adults whose emotions could be regulated by the state. Then they'd have created the new men they desired.

First, this new Soviet man had to be happy. It was his civic duty—responsibility—to love the state and be contented. No matter how hungry he got, no matter how many of his loved ones passed away, his happiness must remain. But this didn't mean grinning like an idiot all the time. Smiling was something American children did; it was how they manipulated their mothers.[26] A new Soviet man would only smile when a smile was required. No manipulation. No lying—which is ironic, given that the whole thing was a lie.

To enforce this bonhomie, an education system was devised that required each age group to join an organization whose primary purpose was teaching kids how they ought to feel. In contrast to America, where *mother* became a synonym for *God* to children, Soviet upbringing was communal. From the start of their lives, Soviet children were taught that the collective is where love comes from. Preschoolers were taught that Lenin was their loving grandfather; young children were instilled with revolutionary zeal in the Octobrist groups—and on it went.

These organizations used ritual, repetition, and emotional displays to instill the right kind of happiness, and they used guilt as a cudgel on children who weren't happy enough. This all might seem to share something with Confucianism—ritualized emotional responses appropriate for particular circumstances. But it didn't. Confucian emotions grew out of the culture that surrounded them. They adapted to changes in culture over time. They were as much bottom-up as top-down. Soviet emotions, by contrast, were entirely top-down, designed and controlled by the state. This was an emotional regime of the strictest kind, and with it came intense emotional labor.

In the Soviet Union, all emotions on public display were interpreted by the state for its own good. The rare times when people might express secret feelings was when they were in private—in places that William Reddy has called emotional refuges. People had to be able to switch their emotions on and off at a moment's notice. They had to be angry at America or happy about their leaders' latest achievements on demand.

People lied about how they felt because if they slipped, they might be diagnosed with one of many Soviet psychiatric ailments, such as "sluggish schizophrenia." The symptoms of this condition included not adapting appropriately (that is, emoting incorrectly), being pessimistic (also emoting incorrectly), and disagreeing with the powers that be (the ultimate in emoting incorrectly). To experience the wrong emotions was to be against the state. To be against the state was to be mentally ill. It must have been exhausting.

The Promised Land

Both sides in the Cold War wanted to control the emotions of their populations and create an affective utopia. They did this to keep their citizens calm in a nuclear age and, perhaps less benevolently, control them. The love the citizens were expected to feel for their respective countries—and in the case of the United States, their mothers—was coupled sharply with a hatred of the enemy.

The Soviets had to have the first satellite in orbit in order to (a) show Soviet citizens that the love they felt for their country was warranted and (b) show the evil Americans that the Soviets were better than they were. America had to get to the moon first for similar reasons: (a) to increase love for the United States among its citizens and allies, (b) to stop the evil Soviets from claiming the moon as their own, and (c) to let the evil Soviets know that Americans were better than they were. But trying to control the emotions of an entire population by force is, historically, a bad idea. Even before Neil Armstrong and Buzz Aldrin planted the Stars and Stripes in the Sea of Tranquility, the cracks were beginning to show.

The problem was that not everyone was considered part of the American way of life. A great many people of color, women, and young men didn't recognize the emotional regime that had been created around them. It didn't take much for these groups to break out from the emotional refuges they'd created through political gatherings and activism and express anger publicly. One match was lit on the evening of April 4, 1968. On that day, Dr. Martin Luther King Jr. was talking to a colleague when a shot rang out. Dr. King died soon after. James Earl Ray, a proud and vocal racist extremist, had killed him. This was a shot that shook the world and unleashed fury across America. The hope experienced by many African Americans and their allies in the civil rights movement was torn asunder. All that was left was rage.

Naked anger exploded in the form of rioting across eleven major cities with a ferocity and level of disruption not seen since the end of the Civil War. This was far from the last riot to shake America. In fact, riots continue to this day when emotional and political regimes imposed on certain groups come face-to-face with the violence used to keep those regimes intact.

It was worse in the Soviet Union, although it took longer for things to unravel. Things really began to fall apart in the 1970s, when the state devised an economic plan it couldn't possibly implement. It didn't help that everyone knew the government was corrupt. Skilled workers were paid too much, and unskilled workers cared too little. A series of economic shocks along with an overwhelmed and understaffed central bureaucracy made matters worse. Stagnation followed, then contraction, crushing the economy.

The USSR never recovered. The heady days of the Space Race became a distant memory—the Soviets had no chance of outspending the United States technologically. Ultimately, the Soviet Union came crashing down like the Berlin Wall. The new Soviet man's unwitting rebellion against state-sponsored emotional control and the citizens' unwillingness to waste any more emotional labor contributed to the collapse.

The Cold War inspired the widespread scientizing and categorizing of emotions as part of an attempt to neatly file feelings. This was a new development in the way the entire world understood emotions, and, as I'll explain in the coming pages, this wasn't universally a good thing, because it led to flawed ideas such as the notion of basic emotions. Furthermore, in the wake of the Cold War people became fixated on finding cultural explanations for human emotions—and completely ignoring evolution—with terrible results. The following chapter will begin to answer the question you might have had for a while now: What do researchers think emotions are today? So let's explore the two main contemporary theories of emotion and discover why both are wrong.

Fourteen

The Great Emotions Face-Off

One bright and sunny morning in July of 2020, a young woman opened her cell phone. She'd been out shopping for the first time in weeks, and, like any good millennial, she chronicled her experience with photos. She scrolled through the many pictures of her shopping trip before deciding on a selfie that seemed flattering enough for Twitter and Instagram. But what to write for the caption? She thought about it and decided she wanted to make a statement. She wanted all 312 of her followers (105 on Instagram) to know how she felt. She wrote, "Out shopping for the first time since March yesterday. Wear a mask, everybody! Be safe!" Then she posted the picture and went about her day.

First, there was one buzz for a response. Then another, then another. Before she knew it, her phone was vibrating so often that it got annoying. She silenced it until her shift at work ended. At the end of the day, she checked it again and was surprised to find 1,513 responses. That couldn't be right. Tentatively, she opened her Twitter feed. Her worst fears were realized. This young lady had been hit by an army of trolls, responded to with a dogpile of shitposts.[1]

Some of the responses looked vaguely like arguments. Masks are fascist; they exist only to oppress; they cause sickness through carbon dioxide poisoning; they hide your face from God. The rest were a mix of comments and memes demonstrating disgust or resentment at her rather mundane post. Some comments bore almost no relation to the topic at hand. Some responses were filled with conspiracy theories. Some were outright insults. These ranged from the harmless, such as calling the young woman a "snowflake" (someone easily offended), to the dangerous, such as threats of rape and murder. And on and on it went. All because she'd suggested that people should wear a mask to protect themselves and others in the middle of a pandemic. She deleted her tweet.

Why people behave this way is something science is attempting to unravel. It turns out that there's a specific emotion—our old friend disgust—at the center of it all. I'm sure that doesn't surprise anyone. What might surprise you, though probably not if you've read this far, is that disgust itself is a complicated and often debated emotion. It's not something universally shared by all humans—well, not exactly. It can also tell us more about people's political views than you might possibly imagine. But we have a way to go before we get to disgust specifically. The bigger issue is that, even after all these millennia of thinking and theorizing, there's still no agreed-upon concept of what exactly emotions are. So let's journey through the debate, and then we can get a little disgusting.

Are You Basic?

In the depth of the Papua New Guinean jungle, a group of men returned to their Okapa District homes from a long day of hunting. Though tribal, these people didn't just hunt and gather for food. Near their settlements, they also cultivated root vegetables to add variety to their food. Even so, they weren't farming as we know it. When the big food sources moved, so did they. These people were also prone to going to war and were known to capture—and sometimes eat—their enemies. This particular community, the Fore, had not changed its way of life for thousands of years. But this day was different.

Something was going to happen that would change their lives, and the history of the science of emotion, forever.

In the center of one of the familial villages, heads turned toward a low, distant rumble. It was like nothing they'd heard before. The noise was accompanied by crashing and cracking as trees fell and branches snapped. Louder and louder, the rumble—almost a growl—came toward them. The people in the village faced the oncoming cacophony, hoping to catch a glimpse of the beast making it. Should they run? Could they catch it and eat it? Around the corner, a strange metal box on wheels came grumbling and moaning up the hill. The box stopped, its side opened, and a group of oddly dressed and strangely pale-skinned humans fell out of the creature's belly. Some of these humans were ridiculously tall, although the one who appeared to be in charge was about their height, albeit whiter-skinned than they ever believed possible. With this pale man was a member of the Fore tribe from another part of the district; they recognized him. The Fore man and the small white person approached them, and the white man spoke in a strange, nasal tongue. His companion translated: "Hello. My name is Paul Ekman, and I'd like to ask you some questions."

Paul Ekman was born in 1934, the child of a pediatrician and an attorney. He spent his youth dreaming of emulating his hero, Ferdinand Magellan, hoping someday to make discoveries that would change the world. But when Ekman was fourteen years old, his mother's depression resulted in her suicide. He felt he had to find out why this happened, why emotions can drive people to end their own lives. His dreams of discovery shifted from geography to the uncharted regions of the mind.

Ekman went on to complete a PhD in psychotherapy, studying the depressed. He was fascinated by nonverbal communication, examining patients' body language and hand movements. Before long, he realized that his patients represented a biased sample: he was studying the survivors of depression, not people who had succumbed to the worst of their illness. He mused that "the road to understanding human behavior and getting back to help people was not by looking at abnormal behavior but at normal behavior."[2] Depression was an emotional disorder, so the man who had idolized Magellan finally found his own quest: to discover if all humans experienced a set of universal emotions.

By the 1960s, Ekman wasn't the only person to have gone searching for basic emotions in tribes that connected us to our hunter-gatherer ancestors. The acclaimed anthropologist Margaret Mead, whom I mentioned in the previous chapter, had already spent years traveling the world, demonstrating that cultures express feelings differently. Most famously, Mead had lived during the 1920s on the small island of Ta'ū, in American Samoa, in an attempt to discover whether the horrors of being a teenager, for parents and kids alike, were universal. What she found was that teenage Samoan girls often engaged in guilt-free casual sex before getting married and beginning a family. This was quite at odds with the norms of the West, where that sort of sex was accompanied by anxiety, shame, embarrassment, and even moral disgust. In 1928, when Mead's *Coming of Age in Samoa* was published, her findings shocked American readers. They provided strong evidence that emotions vary from culture to culture.[3] By the late 1960s, Mead's views were all but scientific consensus in the West, if only as a warning about what could happen if the American way of life and its emotional landscape were abandoned.

Ekman had his doubts. A product of the psychological turn and the new desire to categorize everything, his work likely didn't exist in a vacuum. This is why, by 1964, he was struggling. He couldn't study emotional behaviors without defining them precisely first, but nobody had yet done that. This was when he met psychological theorist Silvan Tomkins, who would go on to become one of his closest collaborators. Ekman found Tomkins's arguments in favor of innate human emotions to be more convincing than Margaret Mead's. He was sure that, if he was going to test his hypothesis, he needed first to figure out a way to measure emotions. The key, he believed, lay in facial expressions. If all humans made the same faces in response to the same feelings, then there must be a link between those facial expressions and innate, basic emotions.

Ekman spent the next eight years alongside Tomkins and another colleague, Wallace Friesen, developing his method. Ekman and Friesen started by approaching students in the United States, Brazil, Chile, Argentina, and Japan and asking them to match photographs of facial expressions with words or stories related to emotions. It quickly became apparent that a basic set of six facial expressions was linked to a basic set of six emotions wherever they looked: happiness, anger, sadness, disgust, surprise, and fear. But Ekman knew there was a problem: all the people tested had access to Western media. They watched American films and TV shows, and they had seen Western art and photography. What they needed were some humans who hadn't seen *I Love Lucy* and had no idea who the Beatles were. The legendary Fore of Papua New Guinea appeared to fill the bill. So the team boarded an old Cessna aircraft and set out to surprise the inhabitants of a Fore village.

On an Island

Ekman and Friesen were careful. They did their best to make sure the members of the tribe they tested hadn't seen any Western media. They even went as far as to make sure they hadn't met any outsiders at all and that they knew not one word of English. They found 189 adults and 130 children who met those criteria. The idea was to use the photos and stories that the researchers had used everywhere else. Ekman and Friesen put their translators through rigorous training in an attempt to make sure that differing translations of the words and stories would not influence the experiment.[4]

Despite having never seen photographs before, never mind photographs of white faces, the Fore caught on quickly. The adults were shown three facial expressions, the children two, and were told a single-sentence story—for example, "This person is about to fight." If emotions were universal, the stories ought to be linked to just one of the pictures. They were. As often as 93 percent of the time, the Fore chose the same matching pairs of stories and expressions as the Western-influenced subjects did. Ekman and Friesen thought they had nailed it: all human beings, everywhere and everywhen, felt those six basic emotions: happiness, anger, sadness, disgust, surprise, and fear.

But there are some big problems with Ekman's idea. First, he and his team weren't the first people to meet and document the Fore. Anthropologists had already studied them, in 1953. In fact, the only reason Ekman's Jeep could reach the Fore at all is that missionaries and government patrols had created a trail for vehicle access. By the time Ekman visited the Fore, the crops they were growing nearby weren't root vegetables but coffee to be sold inland. This meant they were also using money. At one point, Ekman had to pay for a "blessing from the local witch doctor" in order to get things done.[5] In dollars. So much for being untouched by the West.

Another problem was in the way the test was translated. Any translator will tell you that translation is not a matter of swapping the words in one language for the words in another. If you do that, you get gobbledygook. Even words in similar languages can be challenging to match, as we've seen. But the language barrier wasn't the worst of it. The photos did show emotional expressions, but the faces were exaggerated to the point of being silly. They weren't the natural smiles and grimaces most people make unconsciously in day-to-day life. Recent studies by psychologist James Russell and his team have shown that when realistic pictures are used, children don't recognize some emotions until they are as old as eight.[6] Younger kids don't know whether the expression that Ekman labeled as disgust—also known as the gape face—is supposed to be disgust or anger. More recently, a group led by psychologist Lisa Feldman Barrett has found that if you provide people with a wide range of facial expressions in photos and allow them to group the images into categories of their choosing, those categories don't match from one culture to the next (more on this in the next chapter).[7] There's a reason for this—faces are just one aspect of the way we express feelings, and isolated, they're not always enough to go on. Emotions aren't just a face or a voice. They're part of nonverbal communication—body language, if you will. We learn how to use this language from our parents, just as we learn how to use verbal language. Also, facial expressions change their meanings depending on the circumstances. Members of some cultures smile when they are angry or cry when they are happy. To use a personal example, the only times I ever cry and display what Ekman would call a sorrow face is when I'm filled with rage. But it's not just the communication that can change.

Imagine that, without your knowledge or consent, someone pumps you full of noradrenaline, so your heart races, your palms sweat, and your stomach becomes filled with butterflies. In one scenario you are put in a room with an attractive, optimistic person. In another, you are put in a room with a person who is grumpy, irritable, and perhaps less pleasant to look at. It's highly likely you'll feel happy, even excited, in the first room and stressed, perhaps angry, in the second. Two psychologists—Stanley Schachter and Jerome Singer—did this back in 1962, long before Ekman and his faces.[8] It led first to what is known as the Schachter-Singer, or two-factor, theory of emotion and second to stricter ethical regulation of experiments done on humans (or at least one would hope!). Context, it turns out, matters.

Similarly, culture matters. We are taught how we're supposed to behave when we feel something by our upbringing and our culture. I'll go into these factors in detail in the following chapter. But for now, it's enough to know that the universal view of emotion might not be all it's cracked up to be. That didn't stop it from dominating for more than forty years.

Turning Words Around

Twenty years before the fall of the Soviet Union, around the time Ekman was at his intellectual peak, Europe was taking another turn. A linguistic turn. In the 1970s, a host of mostly European thinkers developed a distrust of the Enlightenment. And to explain why, I have to try to describe a very complicated concept as simply as I can. It's not that easy to do, because, to be honest, many of the works that came out of this period are borderline indecipherable.

Basically, some of the finest minds of their generation—philosophers such as Jean-François Lyotard, Jacques Derrida, and Michel Foucault—started to doubt whether the pursuit of truth through science was all it was cracked up to be. They began to wonder if you could know anything for sure. Were race and gender real or constructed by culture? (It turns out they're constructed.) Are those "master narratives" and structures that we take for granted—capitalism, communism, religion, government, even family—real? Or are they just something we've made up? (It turns out they're made up.) Is the modern world just as fleeting as ancient Greece and ancient India? These intellectuals even wondered whether the way we use language is the only way it can be used—hence the unreadable nature of many of their writings. A new way of thinking, often called postmodernism, emerged from these questions. But the linguistic turn included similar notions such as poststructuralism (doubting that intellectual and cultural structures are real) and deconstructionism (taking old ideas apart to determine whether they are real).

This new way of seeing things began in the world of art. Modern artists such as Jackson Pollock were famous for declaring that what they think their art means is what it means, and that's that. The postmodern artists in America were fueled in the 1960s by the rebellions of women, people of color, and the groups we now describe as LGBTQ+. They were also reacting to the emotional status quo of the Cold War era.[9] These artists rejected the modernist stance, proclaiming, "No, I will interpret your art as I see fit, thank you very much. Moreover, you can interpret my art as you see fit."

The core idea was that no single person's view was more "authentic" than anyone else's. For example, suppose we use this notion to study history. It's a terrible fact that Jewish people were rounded up by Nazis and killed in concentration camps. But the individual stories told by the victims who survived aren't facts. They're lived experiences, differing perspectives on a single story. No one story is "truer" than any other, assuming that the person telling it isn't lying. Of course, studying history like that can be very dangerous. In the wrong hands, it could make a Holocaust denier's opinions about the Second World War as relevant as a Holocaust survivor's, and of course they aren't.

Postmodernists also reject the idea that history has a set narrative, goal, or direction of progress. That's why a postmodernist can't be a Marxist. What's interesting is that the concept of modernism has something of a conservative, right-wing tinge to it. Similarly, postmodernism has a somewhat left-wing flavor. But it's not just politics, art, and history that became postmodern. Nothing was immune from the linguistic turn, not even the study of emotions.

Deconstructing Emotions

In 1977, a young American woman sat on a plane flying from Guam to the small island of Yap, in the western Pacific Ocean. For most of her journey, she'd been close to something American—the high-rise hotels of Hawaii, the military bases at Pearl Harbor, the McDonald's restaurants in Guam. She was trying to get past that, to someplace a bit more alien. It was a small plane, and its passengers were a hodgepodge of scientists, government officials, and Peace Corps volunteers. It also carried a few "adventurous Japanese tourists," US naval officers, and construction workers. When she landed in Colonia, America was still with her—gas stations, bars, restaurants, even arts-and-crafts shops. But Colonia wasn't her final destination. She had one more trip to take. After meeting a clan leader named Tamalekar, from the nearby atoll of Ifaluk, she boarded one of the two ships heading for his home.[10]

Ifaluk hadn't been a random choice. Catherine Lutz wanted to spend time with a people whose "gender relations were more egalitarian than in American society,"[11] where women had a louder voice and were free to express their emotions openly. She believed that she had found this in the Ifaluk people. Though not as remote as others, such as the Fore, the Ifaluk had a well-documented, if not well-studied, *otherness* in the way they expressed their emotions. Lutz wanted to know if one culture's emotions could be translated into another's. This meant looking at the same things Ekman was looking at—facial expression, body language, tone of voice, and so on—but seeking differences rather than similarities. She managed to find a few differences, but the one she spent the most time thinking about was *fago*.

At first glance, *fago* seems as if it can be translated as "love" or "compassion," but it's not that simple. *Fago* is a bit more specific than love because it has an element of looking after someone in need. For example, if a person is unwell, you feel *fago*. Not only you, though—the whole community feels *fago* and helps take care of its sick member. The emotion intensifies when someone dies. The *fago* experienced when a child passes away, Lutz recalled, was expressed with wailing, the pounding of chests, screaming, and the flailing of the arms. It wasn't random grief; it was choreographed. People would

take turns coming forward (or rather being invited forward by the closest relatives) to cry in the immediate circle around the body. A careful choreography of grief generally requires that those who are "crying big" (or loudly and deeply) do so closer rather than further from the body, and that those who are not crying move back from it.[12]

I've seen something like this elsewhere. As a young man, I spent some time living in Tunisia. The reaction to the death of a loved one in that culture was much the same and similarly choreographed. Rather than taking turns, each person who entered the home of the deceased was duty-bound to wail louder than the last, to express ever more grief than before in a crescendo of sorrow. What's important to remember is that although it's choreographed, it's not fake. This is the way Tunisians have been taught to grieve. It's the only way they know how to do so. It is entirely natural to them.

So was Lutz just describing grief? Was *fago* just another word for an emotion with which we're all familiar? Well, probably not. *Fago* is both what a mother feels for her child and what a husband feels for his wife when she goes away. It is also felt differently depending on whom it's being felt for. If a person needs you a lot, then you feel a lot of *fago*. If he or she only needs a little help, then you don't feel so much.[13] And that's just the tip of the iceberg. *Fago* is a very complicated feeling, touching every element of Ifaluk life. It's all but impossible to translate in a few words. It's a challenge even using many words. Lutz used more than thirteen thousand of them, and I'm still not sure it's perfectly clear, to an outsider such as I, what *fago* is.

Lutz's research presented a challenge to the idea of universal emotions. She wasn't the first; many anthropologists from Mead onward had made similar claims. But her work took on particular importance within academia. It's an excellent example of the constructionist side of the debate between scholars who think some emotions are innate and scholars who believe emotions are all just made by culture. It's a debate that doesn't seem to be slowing down as people take sides.

Those who tend to believe in things like fighting for social justice—for the political and social equality of sexes, races, and genders, for example—generally have roots firmly planted in the soil of constructivism: the idea that each culture and individual has its own voice, its own way of doing and feeling things. Those who oppose this view are often universalist. They see the emotional world as static and unchanging and anyone who opposes the "norm" as a threat. But there's a twist in this tale. It turns out that, when you zoom in, these two sides aren't very different after all.

Unified by Disgust

Disgust is an example of a great unifier. I wrote about a few different types of disgust above, so I hope I've explained that there isn't one single thing called disgust, even though the feelings you get when you come across something revolting—those yucky, gross, "eeew" feelings—do seem to be shared by all humans (and quite a few animals, for that matter). It stands as a great example of the way emotions can be understood from both the universalist and constructivist standpoints. A universalist scientist such as disgustologist Valerie Curtis will argue that disgust is an obviously evolved safety mechanism helping us avoid harm from pathogens and poisons. A constructivist might argue that the fact that every language has a word for disgust—and that these words mean something almost entirely different in each case—proves that revulsion is constructed. They couldn't both be right, could they? Well, yes, they could. It could easily be the case that there's a basic evolved feeling designed to stop us from poisoning ourselves and picking up parasites. That basic feeling is then adapted, shaped, and manipulated by culture, giving rise to unique variations on a disgusting theme.

But what of those people who responded to the shopper's tweet with comments such as "Masks are fascist," "Masks exist only to oppress," "Masks cause sickness through carbon dioxide poisoning," and "Masks hide your face from God"? In his book *The Anatomy of Disgust*, William Ian Miller noted a particular confession by George Orwell found in *The Road to Wigan Pier*. Orwell, a man who described his upbringing as "lower-upper middle class," was raised to believe that "the lower classes smell."[14] Science backs this up: not that the lower classes smell—they obviously don't—but that people think they do.

Moral psychologist Jonathan Haidt thinks that the more prone an individual is to physical aversion, the more socially conservative he or she will be. He demonstrated this by asking people to complete a test using something called the Disgust Scale (Revised), or DS-R. In the test, people are asked how much they agree or disagree with certain statements on a scale from 0 to 4, 0 being "strongly disagree (very untrue about me)" and 4 being "strongly agree (very true about me)." The statements include: "It bothers me to hear someone clear a throat full of mucous [sic]" and "Even if I was hungry, I would not drink a bowl of my favorite soup if it had been stirred by a used but thoroughly washed flyswatter." It then puts some scenarios to you and asks how disgusting you think they are, again on a scale from 0 to 4. These include: "You see maggots on a piece of meat in an outdoor garbage pail" and "You discover that a friend of yours changes underwear only once a week."[15] As it turns out, people who answer with a lot of 4s tend to be more conservative than those who answer with a lot of 0s.[16]

There are problems with the test. First is the fact that the questions in the DS-R grew out of a predominantly white, middle-class American college culture. The questions in the test feel aimed at Western sensibilities—it seems to assume, for example, that everyone wears underwear or at least has the luxury of changing it every day. Second is the fact that the history of the test shows how aversion can change over even a small amount of time. The original test, the DS, included the question "I think homosexual activities are immoral." It doesn't anymore. Even conservative-minded people have, for the most part, become used to the idea of LGBTQ+ people living and working along-side straight people in the United States.[17] Subsequent tests by other researchers have shown that the idea of disgust, or aversion, is linked as much to purity as it is to right-wing views.[18] In fact, if you are care-ful about what you ask, it seems like people on the far left are just as disgusted by things as people on the far right are.[19]

This is part of the reason why politics in much of the world has become so deeply polarized. Regardless of whether you are left-wing or right-wing, political identity has increasingly come to be defined by a severe reaction to anything seen as "impure." If you are a white male who believes that minorities and women are getting all the breaks, you're likely to view the situation as an infection that needs cleaning up. If you're a feminist racial-equality advocate, LGBTQ+ activist, or even an environmental campaigner, you're likely to see the people who challenge you and all they stand for as contamination, a pimple on the face of equality and justice for all, ripe for squeezing.

The same dynamic underscored the responses to the tweet about wearing a mask. Their authors were disgusted by the young woman's views and took to expressing that disgust in ways that, in turn, were thought of as disgusting. The basic, universal form of revulsion can cause reactions that are poles apart.

As far as the debate between the universalist and constructivist views of emotion is concerned, disgust is the great unifier. It's fascinat-ing, and that's why I love to be surrounded by all things disgusting. Academically speaking, that is.

Turning Inside Out

All this emotion research has had wide-ranging effects. For a start, it has influenced academic study in a number of disciplines. Like stalk-ing giants, Lutz and Ekman have stood over everything that has followed them in their field. Paul Ekman has also affected the world beyond the ivory tower. Since 1978, he has been teaching people to detect what he calls microexpressions—tiny, almost undetectable movements of the face triggered by emotion. He's trained operatives and officers at the CIA, Scotland Yard, and the Department of Homeland Security to spot these microexpressions. Furthermore, I wonder how many of you, kind readers, assumed that there is a set of basic shared human emotions before you picked up this book. If you did, that's thanks to Ekman. Even if you'd never so much as heard his name, his work has been so influential that it's almost taken as a given.

At the same time, the postmodern-influenced constructivist model has had its own impact. It has worked its way into the public consciousness. One size, its proponents argue, does not fit all. This way of viewing the world owes more than a little to Lutz, Mead, and other researchers who first established the idea that we don't all feel the same feelings.

As for disgust, it influences all of us each and every day. We live in an emotional time. For decades we've had emotion-seeking androids, emotionless Vulcans, and emotion-sensitive counselors on popular sci-fi shows such as *Star Trek*. But does disgust drive our political and moral choices today more than it did in the past? That's hard to say. I think the mixing of politics and modern disgust is something new. I suspect it's part of a puritanical worldview that is pushing people apart. Becoming revolted by political actions has become a central part of many Western cultures. And I'm not sure it's for the best.

But as debates about disgust have demonstrated, despite the dominance of the two arguments about emotion in the beginning of the twenty-first century, there is a third way. To explain that in detail, I need to bring us to the present day, perhaps even to tomorrow, and to the cutting edge of science. To the world of artificial intelligence.

Fifteen

Do Humans Dream of Electric Sheep?

It's time to find out what people—or, rather, scientists—think emotions are now. The good news is that the dissolving of the lines in the nature-nurture debate, as disgust so effectively demonstrated in the previous chapter, is becoming ever more prevalent in scientific literature. This is because of the success of thinkers who chose to rebel against the scientific paradigm established by Paul Ekman, Catherine Lutz, and their followers and the failure of a whole field of science— affective computing— as it tries to give real, humanlike emotions to machines. This final chapter will use these successes and failures to explore one of the best theories of emotion that we currently have, beginning with one of the most influential rebels of them all— Professor Lisa Feldman Barrett.

When Science Feels Wrong

In the late 1980s, Lisa Feldman Barrett had a problem. She was running experiments to investigate the way self-perceptions influence emotions, but her results seemed to be inconsistent with what was already published. Eight times in a row, in fact.

At the time, she was a graduate student in psychology at the

University of Waterloo, in Ontario, Canada. As part of her research, she tested some of the textbook assumptions that she had been taught. One of these was the self-discrepancy theory, which I discussed in the context of shame on the shores of Japan (see pages 158–174). According to Edward Tory Higgins, anxiety, depression, fear, and sadness, like shame, spring from the gaps between a person's ideal/ought self and his or her actual self. To test his theory in relation to shame, all Barrett needed to do was ask test subjects to answer questions that would reveal the gaps between people's ideal/ought selves and their actual selves.[1] At the time she was running her experiments, self-discrepancy theory was about as often challenged as the notion that water is H_2O. All she wanted to do was confirm Higgins's hypotheses so she could build on them as she developed her own ideas. But after designing and running her experiments, she discovered that her test subjects weren't distinguishing between feelings of anxiety and depression. They weren't differentiating between feelings of fear and sadness, either.[2]

Barrett wondered why her test subjects were experiencing emotions in a way that didn't fit with what she'd been taught. At first, she thought she had made some kind of grave mistake, chastising herself for developing a faulty experiment. But as she looked through her data, she realized there wasn't anything wrong with her study. The mistake wasn't in who was doing the tests—it was in the way emotions were being measured.[3] She soon found other studies—conducted in the lab and in the field—that also indicated that the ways in which emotions were being measured might be wrong.[4]

Barrett found herself asking an age-old question: If the definition of emotions set forth by Ekman and others is incorrect, then what are emotions? This isn't just an academic exercise. It's central to the question of whether we'll ever build an artificial intelligence that experiences emotions just as we do. She began experimenting and soon constructed a new model, traveling back to the ancient Greeks we began our journey with.

What Barrett hypothesizes is a sort of modern scientific version of emotions as perturbations in the soul. Well, not exactly the soul—more like the body. And not perturbations per se but more like the body being knocked off-kilter. Simply put, she believes that the body tends toward being as energy-efficient as possible. Everything works as it should in safe surroundings, where no harm can come to it. When something comes along to unbalance the body—either something external, such as the threat of a predator, or internal, such as the need to eat—we experience sensations known as core affect. How the brain makes sense of the sensory data that create core affect depends on who we are, where we are, whom we were raised by, and a whole host of other factors. To quote Barrett, "In every waking moment, your brain uses past experience, organized as concepts, to guide your actions and give your sensations meaning. When the concepts involved are emotion concepts, your brain constructs instances of emotion." This is known as the theory of constructed emotion, and I'll explain it in more detail as we go on.

Barrett's work, and that of others who have come to the same conclusion, has changed the game as far as emotion research is concerned. Whereas papers were once dominated by the followers of Paul Ekman or Catherine Lutz, the science of emotions is becoming ever more nuanced, moving beyond just nature or just nurture. To use a personal example, I get an email each afternoon containing a link to every journal article and book about emotions released that day. In the early 2010s, almost every link would take me to either a work about testing basic emotions, or to one defending the constructed emotions of one tribe or another. Now these are an increasing rarity. The field as a whole is slowly shifting toward Barrett's way of thinking.

The changes in the field of emotional history as a result of Barrett's theories can be illustrated by an educational video game some colleagues of mine have been developing called *The Vault*.[5] *The Vault* is a time-traveling puzzle game in which the player wanders through various historical scenarios and solves challenges in order to progress, and the solutions come from understanding how certain emotions were experienced in each section's period of history. The game's premise is that emotions are not static or universal but instead change over time. It tries to immerse the player in unusual, unfamiliar sets of feelings such as acedia, caused by a disconnection from God or the universe. There's also another low feeling, melancholia, which is marked by the sensation of one's body being filled with a horrible black bile and involves an experience of bodily distortion, such as believing one's legs are made of glass.

For me, the interesting thing about this game is that no matter how well I learn to understand historical emotions, I'm not sure I'll ever be able to really feel them in the same organic way that people did in the past. This raises a question: If we humans are unable to fully experience some emotions that were undoubtedly real, such as melancholia, purely because of the historical context in which we live, will machines ever be able to really *feel* anything at all?

This doesn't just apply to historical emotions, either. Ekman may have documented hundreds of facial expressions from cultures around the world, but as I've said, his six basic emotions (inspired by the work of Charles Darwin) came from a small set of American faces, which he then imposed as a framework over the expressions seen in the rest of the world. There's a bias embedded there. Why did he, or anyone else, get to reduce the diversity of human facial expressions, tones of voice, and other behaviors down to any kind of short list?

In the seventeenth century, the standardization of many European languages was often shaped by the arbitrary choices of book printers, who picked one way of spelling a word over another, and this, in turn, shaped cultural expression. Emotion science faces the same problem. Assuming that one set of localized emotional variants will apply to thousands or even millions of people opens the way to a self-fulfilling standardization of emotion rather than to understanding and appreciating the diversity of ways in which people around the world feel, thanks to nuanced cultural differences. Psychologists confront this kind of problem all the time with WEIRDs. That term refers to white educated people from industrialized rich democratic countries—in other words, the typical North American or European undergraduate psychology student, who also happens to be the typical volunteer for a psychological study. This research bias undermines any quest for universal human traits from the start, and the same cultural blind spots are notoriously common in the tech industry.

How to Spot a Paranoid Android

Imagine, if you will, that it's the not-too-distant future. You are lucky because you live on a hill, so the rising tides haven't flooded your hometown yet. You live in a part of the world relatively unaffected by the freakish weather that regularly devastates so much of the planet, and you survived the many wars over water that plagued the middle of the twenty-first century. How do you feel?

You probably feel the same as everybody else. By that, I don't just mean that you share a deep sense of rage at your great-great grandparents for not listening when scientists politely suggested that they burn less fuel and dump less waste. What I mean is that you have the same set of emotions as almost every other human on the planet and express them in nearly identical ways. I realize that this seems to contradict everything I've said in this book so far. I'm aware that I've talked about various kinds of love and hate and era-specific revulsions and angers. I've made an attempt to enthrall you with tales of how honor was lost, found, then lost again. How desire twisted and re-formed to help humanity discover new continents, new science, and fresh takes on religion. How *pathē* became passions, affects, sentiments, and, finally, emotions. How emotions were complicated and cultural, then basic and universal, then complicated and cultural again. We've journeyed through lost passions and examined newly "discovered" emotions. And yet I believe that in the future we will all feel the same. The reason I think this is because slowly, undoubtedly, our feelings are being homogenized by technology. So to wrap up our voyage through history and feelings, let's look toward the future by exploring some of the new technologies that are changing the way we feel. In not entirely positive ways.

The Rise of the Machines

Alan Turing sits alone in a room. In front of him are two slots, each just large enough to pass small notes through. He's been told that behind one slot is a man and behind the other is a woman. His task is to work out who is who by writing questions on pieces of paper before putting them through the slots. What he doesn't know is that behind the first slot, there is indeed a man, but behind the second is not a woman but a computer armed with artificial intelligence.

Turing has already asked quite a few questions, and he thinks he's getting close to solving the puzzle. He writes a new question on two pieces of paper and slips them through the slots: "What did you do this morning?"

The notes return and say mostly similar things about breakfast and catching the train to work, though one emphasizes its kids a bit more. Turing asks another question: "If you close your eyes and remember your childhood, what do you picture?"

The first writes that it pictures playing with toys in its bedroom. It especially liked Flash Gordon rockets and had all the sets. The second remembers playing in the garden, catching butterflies, and wearing ribbons in its hair.

Turing is now sure which is which, but being Alan Turing and having a brain the size of an average football field, he feels like something isn't quite right. The second response was too clichéd, too predictable, too stereotypical. He asks another question: "A little boy shows you his butterfly collection, plus the killing jar. What do you say?"

The first responds: "I'd say, 'That's really interesting, but why don't you keep the killing jar to yourself? It might upset some people.'"

The second writes: "Nothing. I take the boy to the doctor."

Turing is confident he's worked out the ruse. He goes in for the kill:

While walking along in desert sand, you suddenly look down and see a tortoise crawling toward you. You reach down and flip it over onto its back. The tortoise lies there, its belly baking in the hot sun, beating its legs, trying to turn itself over, but it cannot do so without your help. You are not helping. Why?

The first responder answers: "What do you mean I'm not helping? Why would I flip it over in the first place?"

The second answers, "What is a tortoise?"

Alan stands up and declares, "Player 2 is a computer." He wins. The test ends.

Of course, the real Alan Turing never did this. But the World War II code breaker, father of the modern computer, and bona fide genius did come up with a similar test. The goal of the test is to determine whether artificial intelligence has become indistinguishable from human intelligence. If a human can't tell that one of the people being questioned is a computer, the computer passes the test. The real Alan Turing might not have realized that, for his experiment to work, he might have to use Voight-Kampff–style questions, as I did. These questions are used to detect replicant androids in the classic movie *Blade Runner*. I'm sure some of you spotted that. Philip K. Dick was onto something. You'd need to ask questions like these to work out whether something was a machine or not. Because for a computer to fool us into thinking it was human, it would have to trick us into believing it could read and feel emotions. The big question is: Is that possible?

One of the architects of *The Vault*, historian Thomas Dixon, told me that just as "each culture (and each individual) [has its] own different repertoire of feelings," there should be "no reason why an AI machine should not be able to learn those patterns."[6] However, Barrett argues that the facial expressions, vocal tones, and behaviors associated with emotions could well change not only from culture to culture but also, subtly, from person to person.[7] Are we ever going to be able to build a machine that can recognize the full diversity of human emotional experience, even when the differences between two cultures' emotional expressions are so great as to be untranslatable? Or are we going to build one that only recognizes a small sliver of that diversity and so, accidentally or otherwise, forces its users to change their behavior to match?

Computers can do things much more quickly than humans. For example, they can instantaneously search databases for millions of pieces of information that it would take a human years to find. But computers can only do what they're told. When Alan Turing wanted to break a German code using a 1940s-era computer, he had to put all the data about that day's code into the machine first. He then had to examine the readouts and decide which parts were useless and which his superiors needed to know about as soon as possible. His computer couldn't intercept and process the coded German transmissions on its own, and it couldn't work out which messages were important. The same is essentially true of all modern computers—that is, it was true until the twenty-first century.

A computer equipped with artificial intelligence gives time back to humans. AI can read millions of documents and, crucially, learn things about those documents in the process. For example, in 2020, AI systems were set up in East Africa by the University of Cambridge and the Infectious Diseases Institute, in Uganda. Their job is to read the handwritten medical records of the many millions of HIV patients in the region and spot patterns in factors such as economic background, geographical location, tribal affiliation, age, and so on. Almost instantly, AI can highlight the traits that cause some people to miss out on medicines while others get the drugs they need. Humans could recognize these patterns, too—in fact, they have been for years—but it takes ages. This sort of AI can save lives, but it doesn't need emotions to do so. The deeper reasons why some people aren't getting their medicine is still beyond basic AI, and that's where emotions come in.

It's impossible to make many types of decisions without feelings. For example, imagine you're choosing which flavor of ice cream to buy. This isn't a decision you can make using pure logic. You choose an ice cream flavor by recalling the pleasure created by the taste of the ice cream (and perhaps other foods) you've had in the past. A purely logical android could never choose. And it certainly wouldn't have a favorite flavor.

We don't just use emotions to choose desserts, though. Neuroscientist Antonio Damasio has shown that almost every decision we make taps into our memories of good and bad things.[8] If we did something in the past and it went badly, we don't do it again. If it went well, we do it again. This applies to anything from saying "yuck" to garlic-flavored ice cream to deciding which seat to take on the bus. Ask yourself why you sit where you sit on a bus, train, or plane. Without emotions, fuzzy, nonbinary decisions are almost impossible to make—you'd be stuck looking at seats forever.

Despite these problems, there's an entire field dedicated to creating emotional machines: *affective computing*. The term was coined by computer scientist Rosalind W. Picard in 1995, and she remains an active researcher today. Her field of study tries to do two things. First, it works to create machines that recognize feelings. Second, it seeks to develop computers that can feel emotions themselves. The latter is a rabbit hole down which I shall not descend here, but the first goal of affective computing, creating machines that recognize human emotions, is one of the reasons we'll all feel the same as everyone else does in 2084—assuming, that is, that people like Dr. Picard achieve their goals.

Big Brother, 2084

Most of the work in affective computing has gone not into building robots but into finding ways to use emotion-recognition technology to sell you things and perhaps increase your safety and happiness as well. Technology you likely have in your home can already do this. Nearly one-fifth of American adults own an Amazon Echo, Google Home, or equivalent smart speaker. Amazon wants people to trust its virtual assistant, Alexa, which uses whispers, shouts, and variations in pitch and speed to indicate emotions and make itself sound more "human." It's also been programmed with so-called delighters—random humanizing responses, such as terrible jokes, beatboxing, and silly songs. The Google Home speaker does similar things—if you have one, try telling it that you think Alexa is better. The response I got was, "Well, we all have our opinion," followed by audible weeping. Amazon, Google, and other tech companies hope that this sort of emotional expression will make their technology feel more realistic. They hope to give their products a lifelike essence, create stronger bonds, and therefore retain more loyal customers.

These companies also expect to use affective computing technology to monitor—and, ideally, improve—the mental health of their customers. Amazon is already developing software that analyzes our voices in order to detect moods. When you get annoyed, Alexa will calm you down. When you're happy, it can join in your joy. If you appear down, even suicidal, Alexa can play a wellness or meditation program to get you through. This is no bad thing in principle. Evidence mounts that mental health issues, even thoughts of suicide, can be alleviated by conversation, including conversation with AI chatbots.[9]

Emotion-detecting technology is also being developed to protect us while driving. For example, a company called Affectiva wants to monitor drivers, identifying the emotions in their voices, body language, and facial expressions.[10] If you get a bad case of road rage or look like you're feeling blue to the point of driving dangerously, the company hopes its automotive AI will be able to take control of the car, take you to the nearest safe place, and give you time to collect yourself before you carry on or call for help.

This sort of facial-emotion analysis has some other fairly controversial uses, such as deciding how attractive we are. Some researchers have been using AI to try to determine which nose job is less likely to make people look ugly when they make certain faces. Does a petite nose make crying less hideous? Can a less flared nostril take away the unsightliness of the disgusted gape face? Maybe a pointier nose will make your smile more appealing. Computers are being developed to build you a perfect face, free from the ugliness of a bad mood.[11] I can imagine that by 2084, rich people will become almost indistinguishable from the robots that build their new noses. That's assuming they are white and Western, of course.

Affective computing technology is also being deployed as a crimefighting tool, with some pretty bleak consequences. As I mentioned in the previous chapter, Paul Ekman has been teaching people — namely, operatives and officers at the CIA, Scotland Yard, the Department of Homeland Security, and other agencies — to detect microexpressions since 1978. In most cases, it hasn't gone well. For example, in 2007, the TSA launched a program called Screening of Passengers by Observation Techniques, or SPOT. Airport security officers were trained to read people's microexpressions while they waited for their planes. The idea was that no matter how well trained and focused a suicide bomber might be, tiny microexpressions of badness would give him away. Of course, because emotions don't really work that way, the whole program was a complete failure. If you've ever been to an airport, you know that they aren't exactly bastions of calm. The stress of flying makes passengers look and act in weird ways. Faces that reflect badness are everywhere. Moreover, Ekman will be the first to tell you that microexpressions, if they exist at all, are often invisible to the human eye. The TSA's results were abysmal—they would've done better if they'd detained people randomly—and quite a few innocent people missed flights, were arrested, and were denied entry to the United States for no reason at all. It wasn't Ekman's or the TSA's finest hour.

But where humans fail, technology tries to pick up the slack. Researchers at the University of Rochester, in New York State, have crowdsourced photos of more than a million faces to build a database of microexpressions.[12] They want to train computers to do what humans couldn't: assess whether someone waiting in line at an airport might be a terrorist based on his or her facial expressions. Gone are fallible human brains, replaced by emotion-detecting AI that watches people in airports through CCTV and in police interview rooms. A robot bad-guy detector.

This technology doesn't work very well—because you can't identify the emotions of individual human beings by analyzing their microexpressions—but that hasn't stopped what ought to be science fiction from becoming reality. For example, since 2018, the sunglasses worn by some Chinese police officers have had facial-recognition technology built into them. Emotion detection won't be far behind.[13] Western, Ekmanized emotion detection, that is. And that's a problem.

The idea that all humans feel a uniform set of six emotions has made its way deep into the world's collective consciousness. Disney even made a popular animated film in which personifications of five of the six basic emotions were the main characters: *Inside Out*.[14] We are more likely to alter our behavior because of technology than technology is to adapt. The truth is, mapping real human emotions in all their complexity might not even be possible with current technology. Worse still, developing systems that can track the fuzzy reality of emotion based on core affects, language, cultural understandings, context, memories, individual differences, and so on would be impossibly expensive. Money talks.

If governments think that the expensive technology they bought can identify criminals by their unpredictable behavior, people will simply start acting in predictable ways to avoid getting into trouble. The Western ideas of emotion that underlie this technology will ride roughshod over cultural differences, and it won't matter. The die has been cast, and the technology is Western—as all emotions soon will be. That will be a shame, because as I hope you've realized, the vast diversity of ways to understand and express emotion is a wonderful thing. For humanity to take another step closer to uniformity in an area so diverse, so fundamentally human, would be a great loss. I also fear it won't make the world a better or safer place if emotional misunderstandings, and our ability to recognize and acknowledge them in one another, eventually disappear.

Computer Core Affect

Affective computers don't just use facial expressions to detect how we feel, though. They also listen to our voices. The software analyzes the pitch and speed of the voice as well as qualities such as breathiness, loudness, and how often we pause. Some have even gone as far as to monitor body language, mapping gestures onto algorithms to identify our moods. Most modern systems use all three techniques. But there's a problem, one you've probably already guessed: these systems don't work, mostly because they don't take several important things into account.

After years of research, Lisa Feldman Barrett has concluded that emotions are much more complicated than faces and voices. She, along with James Russell, helped develop a more nuanced system than Ekman's. They call it the "psychological construction of emotions" model.[15] It posits that emotions are "constructed" when the brain processes a number of psychological factors simultaneously—internal feelings, perceptions of what's going on in the outside world, patterns that individuals have learned throughout their lives from family and culture, and so on.

Similarly, our brains recognize emotions in other people by observing their bodily or facial movements and, crucially, the *contexts* in which those gestures and expressions are made. When it comes to the question of whether AI can ever experience emotion, it's the complexity of emotion—such as understanding the context, cultural and otherwise, in which emotional expressions are made—that proves the undoing of the machines.

Emotions in Context

Look at this picture of my face and arm. Do I have road rage? Or am I celebrating hearing on the radio that my team has scored?

Even humans struggle to figure this kind of thing out. If we want to put emotion-detecting AI in self-driving cars so they can take control and pull over when they detect road rage, my celebratory fist pump—

not that my team scores *that* often—could mean I end up spending time stuck on the side of the street getting angrier and angrier at my annoyingly "intelligent" car.[16]

To avoid roads jammed with furious drivers, the ability of emotion-processing AI to understand context is essential—and context requires recognizing something that emotion scientists call *value*. This is the meaning that we construct regarding the world around us. If you see me pumping my fist, my value to you depends on whether you thought I was violent, in which case my value is as a threat; whether you knew of my fabled physical cowardice, in which case my value to you is as a joke; whether you were watching the game with me and supported the same team, in which case my value to you is as a friend, or a different team, in which case my value to you is, possibly, as an enemy; and so forth.

But understanding value isn't as easy as building a database of all these factors. As Barrett explains, "Brains don't work like a file system. Memories aren't retrieved like files. Memories are dynamically constructed in the moment. And brains have an amazing capacity to kind of combine bits and pieces of the past in novel ways."[17] Our brains seem to use disparate feelings and memories to build a framework of categories for understanding context; these categories are then filtered and distorted in ways that help us react appropriately to new experiences. This is one of the key reasons why eyewitness testimony is often unreliable in court and cross-examination is an essential part of the legal process.[18]

For a machine to understand the emotions I'm feeling during my fist pump, it needs to contextualize all this: memories of fist pumps and scowling faces, memories of what a car is, memories of various sports and reactions to sports, memories of how rarely my team scores, an understanding of how I feel about my team, an analysis of my driving, a recognition that those are tears of joy rather than sorrow (or anger), and so on. The sum total of this mishmash? An emotion.

Machines may remember things perfectly, but the human emotional process only works *because* it's so fuzzy—it makes cognitive dissonance possible. A machine might recognize my scowl and fist as a threat but simultaneously know that I'm not a violent person. Which information should it react to? The brain can do this with hundreds of bits of conflicting data, and we usually end up able to handle new contexts that might seem logically impossible to an AI. My car might not just pull over—it could shut down completely in the middle of the highway. Then I'd really be angry.

The idea that our memories aren't simply recording devices but rather classification systems that help us thrive and survive is known as *dynamic categorization*. It's a model that's now taken as a given within the field of psychology, and researchers in other nonscientific fields, such as history, also use it. Forcing AI to access memories chosen for it would therefore impose a narrow set of WEIRD-like values onto this hypothetical feeling machine, but developing an AI that can create its own humanlike memories and values as it learns would avoid this pitfall. However, as far I'm aware, there don't appear to be AI or computer systems in development that use dynamic categorization when thinking about memory, let alone emotion detection.

Calculating Emotions

Let's say we solve the two major problems I've raised: our AI device can recognize faces, voices, and behavior, and it uses dynamic categorization to store and recall information. What we've built is a machine that only *recognizes* emotions. It's a metal psychopath—it can't *empathize* with me. I, for one, don't like the idea of being driven around by one of those machines.

So our final step in building a feeling machine is, well, to introduce *feelings*. The ability to understand the value of the world around an organism did not evolve separately from the senses that let that organism learn about the world in the first place. Without feelings of revulsion caused by smell and taste, we all would have died from eating rotten food long ago. Without hunger, we'd starve. Without desire, we wouldn't reproduce. Without panic, we might run toward saber-toothed tigers, not away from them. And just as our sensory abilities extend far beyond the five traditional senses, we also have a wide and diverse range of internal feelings, known in psychology as affects.

Affects aren't emotions but rather judgments of value that create pleasant or unpleasant sensations in the body, either making us excited or calming us down. Affects help us evaluate context, telling us whether we're seeing a dog we can trust (and domesticate) or a tiger from which we should flee. To feel affects, our AI needs a body. As Barrett has argued, "A disembodied brain has no bodily systems to balance; it has no bodily sensations to make sense of. A disembodied brain would not experience emotion."[19]

A feeling machine's body doesn't have to be a *Blade Runner*-style flesh and-blood replica of a human. It could be a virtual body, built entirely from lines of code but serving the function of a physical body as far as the digital brain attached to it is concerned. Unfortunately, most AI developers who imbue their creations with an understanding of emotion—even those who connect their creation to a body of some kind—are building machines that, at best, react in simple ways to basic stimuli such as sight, sound, and pressure. More, much more, is needed to truly create a feeling machine. That said, building such an AI would be an effective way to figure out what that "much more" is. Right now, we have only a few clunky methods for testing our ideas about emotion on humans: we can get people to take a survey; we can put them in loud, claustrophobia-inducing metal tubes and ask them to "feel emotions as they would naturally"; or we can study the effects of physical changes in the brain, whether resulting from an accident or as a side effect of surgery. When it comes to understanding emotions directly, we're still fumbling in the dark. A feeling machine could turn on the light.

As a theoretical exercise, we can imagine that we've cracked it. We've done the experiments, and we've created a machine that can experience affects, read context, and understand value—and all those abilities have been synced up perfectly in order to construct emotions. Plus, we've managed not to imbue our creation with our own cultural biases. We have an emotional, feeling machine. There's one final issue. Creating a machine that experiences emotions doesn't tell us whether we have a machine that *feels* emotions in the same way we do. It might act as if it does, and it might say that it does, but can we ever truly know that it does?

Let's go back to the example of the self-driving, road-rage-detecting car. Sure, it might understand why I'm scowling and raising my fist. It might appear to empathize with me—it understands the value of my gesture. But what we've got here is just a kind of clockwork approximation of something that feels empathy, and we still don't fully understand how all the gears mesh together to create the "feeling" of an emotion. For all we know, the ability to feel may still prove impossible to artificially create. Perhaps that doesn't matter. After all, I can't even be sure whether you, the human reader of this book, really feel emotions in the way that I do. I can't climb inside your brain. Barrett argues that emotions depend not only on a human mind's perception of its own affects, contexts, and values but also on how those perceptions work "in conjunction with other human minds."[20] If we—or other AI minds, even—think that a machine can feel, is that enough?

This question raises another: Is it important that something—a machine, a person—actually *feels* emotions? Or does it just need to behave *as if* it feels emotions? We treat animals differently based on their levels of intelligence, for example, but where does emotion come into play when we're trying to draw the line between a machine and a creation that is artificial but alive? The distinction between AI that only *appears* terrified of death and AI that is *genuinely afraid* is an important one if we ever have to turn the machine off. Can we even make that decision if there's ambiguity?

To come back to the question that I asked back in the introduction: What are emotions? I'm ready to give you my answer. To me, they are the way we use the sum of our experiences to understand how we feel in particular circumstances. Who you are, where you are, and what you are doing are just as important as what you feel. Each language has its own words for feelings. Each culture, even each family, has its own understanding of how you should behave when experiencing those feelings. In the end, even though we might all share neurochemistry that produces similar core affects that evolved to keep us alive—such as the urges to fight and flee—the way we psychologically construct meanings from those feelings is different for each of us. One person's disgust is tied to God. Another's is linked to food. One group might feel that hatred for another group is justified. Another might experience a flash of that same hatred but extinguish it beneath more powerful sensations of shame. One couple's love might be simple, pure attraction; another's might be a complex of emotions, all acting simultaneously to bring them together.

This hugely complicated construct might seem beyond the capabilities of AI right now, but researchers are working on it. Ultimately, whether we're ever able to produce machines that experience emotion may depend not on the skills of scientists but on the ethical considerations of ordinary people. I have few suggestions on this front—but if I were contemplating a career in philosophy right now, I'd be thinking about making my central field the ethics of AI with regard to emotion. There's much work to be done.

Epilogue

The Last Feelings?

When we began exploring the history of emotions together, I wanted to demonstrate three things: first, that emotions aren't universal but in fact change from culture to culture and from historical period to historical period. I hope that's become obvious. Although we may all share some sort of primary set of internal feelings, the way each culture understands the much more complicated experiences that constitute *emotion* is different. Most of us don't have a separate category of feelings called sentiments, reserved for behaving in the right way around—for example—an objet d'art. Or a category for the feelings we get when we detect ripples in our souls, called passions or *pathē*. Even those seemingly ubiquitous emotions that in English we label desire, disgust, love, fear, shame, and anger aren't as universal as many people might assume. Subtle changes in the way cultures understand their feelings can create new religious beliefs and new economic superpowers.

My second goal was to show that emotions are more complicated than you might think. Emotions, it's clear, aren't just some stimulus to the brain followed by a response. They aren't just a face we make or a sound we utter. They are rich and complex, operating on many levels and changing from circumstance to circumstance. The desire that drove humanity to find the New World is not the same as the desire to reach *nirvana*. The revulsion that helped fuel the witch crazes isn't the same as the disgust understood by modern psychologists. While there might be some common, evolved experience, some core affects that humanity and perhaps even all mammals experience, I hope it's now evident that there's much more to it than that.

Finally, I wanted to show you that emotions have a history. This book contains quite a lot of history told in quite a lot of ways: The history of what people thought feelings were and what they think they are; the history of how emotions influenced world events and how the manipulation of emotions can lead to good or bad things. The history of the understanding of emotional disorders and the ways that understanding has changed dramatically over time. Accounts of the ways people reacted to shame, what their gods might have felt in response, and how we should control our feelings.

I said this book would explain how emotions built the world. Each of these chapters, I hope, has served as an example of a world being built in no small part because of the emotions of the people living in it. We saw how emotions influenced the construction of the great religions. We saw how desire helped build the nation-state and, with it, capitalism. We saw how feelings not only the influenced the creation of the United States but also gave rise to the drive to discover new continents. We saw how emotions supported the rise of modern Japan through an act of shame and modern Ghana through a moment of rage. We even saw how emotions are shaping science and technology right now, at this very moment. What's great about that last sentence is that I can write it with the confidence that it will never need an update in a future edition. Emotion will always be shaping our science and technology, now and in the future.

But what about that future? What might emotions look like in the future? That is a long and complicated tale in itself that deserves its own study. But let me give you a little glimpse of my perception of the future of emotions.

Emoticonsciousness

In an alternative universe, in the year 1969, Neil Armstrong finds himself on the bottom rung of a ladder that descends from the lunar module *Eagle*. In this universe, the text message was invented long before the radio. So instead of making a speech, Armstrong sends the world a message:

👌 ⇔ 🚶 👤 👌 ⬥ 🚶🌍[1]

These are *emojis*, and in our universe, they are becoming increasingly common in social media, blogs, text messages, and even marketing communications. One of the most exciting things about emojis and their parents, *emoticons*, at least for those of us interested in linguistics, is that they herald the birth of a new written language. This new language is being created in record time from the ground up by the people using it. And what's intriguing about emojis is that this new language seems to be international. Emojis are increasingly understood by everyone, everywhere, in a uniform way.

It hasn't always been that way. At first, emoticons were slow to take off. When Scott Fahlman first put a message on an online bulletin board that suggested using :-) for jokes and :-(for serious communications, it was still 1982. Most people didn't even know what an internet was, and the World Wide Web was still a glint in Tim Berners-Lee's eye. It took a while for technology to catch up. But when it did, these tiny expressions of emotion, previously the exclusive domain of computer geeks and cybernerds, took off.

What's interesting about the original emoticons is that at first, they told us something about cultural differences in emotional expression. For example, Westerners tend to write emoticons as Fahlman did: oriented horizontally and focusing on the mouth—e.g., :-) for happy and :-(for sad. Japanese emoticons—known as *kaomoji*—are oriented vertically and concentrate on the eyes: ^_^ for happy, '_' for sad. Do people from Asia focus on the eyes and Westerners on the mouth when reading emotions? Some studies seem to suggest so. This is more evidence of the idea that a smile means "happy" to everyone, everywhere, everywhen, is wrong.[2]

There's also evidence that people of various ages read emojis and emoticons differently. How quickly you work out what the other-universe Armstrong texted to us gives away your age.

All this is starting to change, though. Emojis are beginning to replace emoticons completely. At the time of this writing, there are some differences in emojis across cultures that are carried over from emoticons. For example, "sad" is ☹ in Western countries and ☺ in Japan. But these images are also starting to merge: ☺ means "crying" in both Western and Eastern traditions, and "happy" is a version of ☺ in both. Many linguists who study this area seem to believe that emojis are forming a new international language and new global expressions of emotion with it.[3] By 2084, there is a good chance that the whole world will be expressing its basic emotions as ☺, ☺, ☺, ☺, and ☺.[4]

With both affective computers and emojis pushing everyone toward a homogenous experience of emotion, I think that all emotional expressions will start to look identical over the course of the fifty years following this book's publication. Ekman's much-debunked basic emotions are becoming universal as Western technology and its assumptions about basic emotions spread through more of the world. Deviations from them will soon, at best, stop you from being correctly understood on social media and, at worst, land you in jail. The many and diverse ways of emotional expression, which have varied from culture to culture and from period to period, are merging into one. The rich tapestry of emotion found across the real world is disintegrating into the digital world around us.

This is a long and complicated topic, and to go any deeper will take a bit more time than I have here.[5] But I do think the last hurrah of

Western imperialism will occur when it accidentally makes the whole world's emotions Western. And that is why, in 2084, we'll all feel the same. Ekman's basic emotions will have become universal after all.

Final Thoughts

In my opinion, you can't have history without emotion. I hope I've demonstrated that throughout this book. And I hope I've at least shown you that emotions are an essential part of knowing the past. After all, to study history is to attempt to know the past, and how can we do that if we don't try to understand how people felt?

My primary goal was always to open your mind to the history of emotion, to offer a jumping-off point from which you can explore a whole new world of history. There are dozens of books out there on the topic that range from casual narratives to dense analyses (see the list on page 291). Some are about feelings and emotions in general; others focus on specific emotions and behaviors such as weeping, fear, happiness, even curiosity. I hope I've gotten you interested enough to dip into them. I encourage you to seek them out and become as excited about this topic as I am, because I believe that without a history of emotion, we really have no history at all.

Acknowledgments

Writing this book has been challenging, rewarding, frustrating, elevating, and, most important, life-changing. I'm pretty sure I couldn't have done it without a lot of support, not least from my amazing wife and fellow life traveler, Dawn Firth-Godbehere. You are my rock, my night and my day.

I'd also like to thank my family—my mom, Pauline Hart, along with my brothers, Peter, Andrew, and David, and my sister, Jayne, who have been endlessly supportive no matter what mad direction my life takes.

I also have to thank my agent, Ben Dunn, who read an article online that was causing trouble in the AI community and thought, "This guy's got a book or two in him." It was a bold decision. Thanks are also due to my US editor, Ian Straus, whose extraordinary patience with this first-time writer helped shape a book of which I'm rather proud.

Last but not least, I want to thank the people responsible for getting me to the point where I could do something like this—my PhD supervisors, Professor Thomas Dixon and Dr. Elena Carrera; all the people at the Centre for the History of the Emotions, Queen Mary University of London; the Wellcome Trust, for supporting my early disgusting research; and finally, my onetime undergraduate tutor and now friend, Sarah Lambert, a scholar of unshakable intellect who does not suffer fools gladly but somehow gladly suffers me.

Thank you all.

Suggested Further Reading
(to Get You Started)

Fay Bound Alberti: *A Biography of Loneliness*
Rob Boddice: *The History of Emotions*
———: *A History of Feelings*
———: *The Science of Sympathy*
Elena Carrera: *Emotions and Health, 1200–1700*
Thomas Dixon: *From Passions to Emotions*
———: *Weeping Britannia*
Stephanie Downes, Sally Holloway, and Sarah Randles, eds.: *Feeling Things*
Ute Frevert: *Emotions in History — Lost and Found*
Ute Frevert, ed.: *Emotional Lexicons*
Ute Frevert et al.: *Learning How to Feel*
Daniel M. Gross: *The Secret History of Emotion*
Sally Holloway: *The Game of Love in Georgian England*
Colin Jones: *The Smile Revolution*
Robert A. Kaster: *Emotion, Restraint, and Community in Ancient Rome*
Joel Marks and Roger T. Ames, eds.: *Emotions in Asian Thought*
Dolores Martín-Moruno and Beatriz Pichel, eds.: *Emotional Bodies*
Charlotte-Rose Millar: *Witchcraft, the Devil, and Emotions in Early Modern England*
Keith Oatley: *Emotions*
Gail Kern Paster, Katherine Rowe, and Mary Floyd-Wilson, eds.: *Reading the Early Modern Passions*
William M. Reddy: *The Navigation of Feeling*
Barbara H. Rosenwein: *Anger*
———: *Emotional Communities in the Early Middle Ages*
———: *Generations of Feeling*
Tiffany Watt Smith: *On Flinching*
———: *Schadenfreude*
———: *The Book of Human Emotions*
David Houston Wood: *Time, Narrative, and Emotion in Early Modern England*

And a shout-out and apology to the hundreds I've missed.

Notes

Introduction: *How Do You Feel?*

1. Thomas Dixon, *From Passions to Emotions: The Creation of a Secular Psychological Category* (Cambridge, UK: Cambridge University Press, 2003).
2. Anna Wierzbicka, *Imprisoned in English: The Hazards of English as a Default Language* (Oxford, UK: Oxford University Press, 2013), 75.
3. Debi Roberson et al., "Colour Categories and Category Acquisition in Himba and English," in *Progress in Colour Studies*, vol. 2, *Psychological Aspects*, ed. Nicola Pitchford and Carole P. Biggam (Amsterdam: John Benjamins, 2006), 159–72.
4. Jonathan Winawer et al., "Russian Blues Reveal Effects of Language on Color Discrimination," *Proceedings of the National Academy of Sciences of the United States of America* 104, no. 19 (May 8, 2007): 7780–85.
5. Two great examples are Thomas Dixon, *Weeping Britannia: Portrait of a Nation in Tears* (Oxford, UK: Oxford University Press, 2015); Joanna Bourke, *Fear: A Cultural History* (London: Virago Press, 2006).
6. Check out the second half of William Reddy's *The Navigation of Feeling: A Framework for the History of Emotions* (Cambridge, UK: Cambridge University Press, 2001). For that matter, check out the first half, too.
7. A great example of this is Stephanie Downes, Sally Holloway, and Sarah Randles, eds., *Feeling Things: Objects and Emotions Through History* (Oxford, UK: Oxford University Press, 2018).
8. Richard Firth-Godbehere, "Naming and Understanding the Opposites of Desire: A Prehistory of Disgust 1598–1755" (PhD diss., University of London, 2018), https://qmro.qmul.ac.uk/xmlui/handle/123456789/39749?show=full.
9. See Reddy, *The Navigation of Feeling*. In this field, it is a must-read.
10. Arlie Russell Hochschild, *The Managed Heart: Commercialization of Human Feeling* (Berkeley: University of California Press, 1983), 7.
11. Barbara H. Rosenwein, *Emotional Communities in the Early Middle Ages* (Ithaca, NY: Cornell University Press, 2007).

Chapter One: *Classical Virtue Signaling*

1. Plato, "Phaedo," trans. G. M. A. Grube, in *Plato: Complete Works*, ed. John M. Cooper (Indianapolis: Hackett Publishing, 1997), loc. 1792, Kindle.

2. David Sedley, *Plato's* Cratylus, Cambridge Studies in the Dialogues of Plato (Cambridge, UK: Cambridge University Press, 2003), 10.

3. Alfred North Whitehead, *Process and Reality*, ed. David Ray Griffin and Donald W. Sherburne (New York: Free Press, 1978), 39.

4. Plato, "Republic," trans. G. M. A. Grube, rev. C. D. C. Reeve, in *Plato: Complete Works*, ed. John M. Cooper (Indianapolis: Hackett Publishing, 1997), loc. 26028–27301, Kindle.

5. Plato, "Republic," loc. 27176–95.

6. Plato, "Republic," loc. 27239–64.

7. Xenophon, *Memorabilia*, trans. Amy L. Bonnette (Ithaca, NY: Cornell University Press, 2014), loc. 514–20, Kindle.

8. Plato, "Phaedo," loc. 2889.

9. Plato, "Phaedo," loc. 2886.

10. Plato, "Phaedo," loc. 2878–97.

11. Plato, "Critias," trans. D. Clay, in *Plato: Complete Works*, ed. John M. Cooper (Indianapolis: Hackett Publishing, 1997), loc. 1541, Kindle; Plato, "Phaedo," loc. 2511.

12. Plato, "Phaedo," loc. 2890.

13. Emily Wilson, *The Death of Socrates* (Cambridge, MA: Harvard University Press, 2007), 114.

14. Or in the original German, "Oh Kriton, das Leben ist eine Krankheit!" See Friedrich Nietzsche, *Die Fröhliche Wissenschaft*, NietzscheSource.org, http://www.nietzschesource.org/#eKGWB/FW-340.

15. Glenn W. Most, "A Cock for Asclepius," *Classical Quarterly* 43, no. 1 (1993): 96–111.

16. Xenophon, *Memorabilia*, loc. 2859.

17. Plutarch, *Plutarch's Lives*, trans. George Long and Aubrey Stewart (London: George Ball and Sons, 1892), 3:302.

18. A good argument for this is found in Bente Kiilerich, "The Head Posture of Alexander the Great," *Acta ad archaeologiam et artium historiam pertinentia* 29 (2017): 12–23.

19. Pseudo-Callisthenes, *The Romance of Alexander the Great by Pseudo-Callisthenes*, trans. Albert Mugrdich Wolohojian (New York: Columbia University Press, 1969), 57.

20. Aristotle, "On the Soul," trans. J. A. Smith, in *The Complete Works of Aristotle*, ed. Jonathan Barnes (Princeton, NJ: Princeton University Press, 1984), 1:413a20.

21. Aristotle, "On the Soul," 1:434a22–434b1; Aristotle, "Parts of Animals," trans. W. Ogle, in *The Complete Works of Aristotle*, 1:687a24–690a10; Aristotle, "Metaphysics," trans. W. D. Ross, in *The Complete Works of Aristotle*, 2:1075a16–25.

22. Aristotle, "On the Soul," 1:424b22–425a13.

23. Aristotle, "On the Soul," 1:1369b33.

24. Aristotle, "On the Soul," 1:1370a1.

25. Aristotle, "On the Soul," 2:1378a30, 1380a5.

26. Aristotle, "On the Soul," 2:1380b35, 1382b1.

27. Aristotle, "On the Soul," 2:1382a21, 1383a15.

28. Aristotle, "On the Soul," 2:1383b16–17.

29. Aristotle, "On the Soul," 2:1385a15–1385a1.

30. Aristotle, "On the Soul," 2:1385b13, 1386b10.
31. Aristotle, "On the Soul," 2:1387b20,1388a30.
32. Aristotle, "On the Soul," 58.
33. Aristotle, "On the Soul," 2:1379b1–2.
34. Aristotle, "Rhetoric," trans. W. Rhys Roberts, in *The Complete Works of Aristotle*, 1:1367b8.
35. Pseudo-Callisthenes, 59–60.
36. Pseudo-Callisthenes, 59–60.

Chapter Two: *Indian Desires*

1. Harry G. Frankfurt, *On Bullshit* (Princeton, NJ: Princeton University Press, 2005); Harry G. Frankfurt, "Freedom of the Will and the Concept of a Person," *Journal of Philosophy* 68, no. 1 (January 14, 1971): 5–20.
2. Timothy Schroeder, *Three Faces of Desire* (Oxford, UK: Oxford University Press, 2004).
3. Wendy Doniger, *The Hindus: An Alternative History*, reprint ed. (Oxford, UK: Oxford University Press, 2010), 44.
4. Upinder Singh, *A History of Ancient and Early Medieval India: From the Stone Age to the 12th Century* (New Delhi: Longman, an imprint of Pearson Education, 2009), 19; K. S. Ramachandran, "Mahabharata: Myth and Reality," in *Delhi: Ancient History*, ed. Upinder Singh (Oxford, UK: Berghahn Books, 2006), 85–86.
5. Eknath Easwaran, trans., *The Bhagavad Gita* (Tomales, CA: Nilgiri Press, 2007), 251–65.
6. Daya Krishna, "The Myth of the Puruṣarthas," in *Theory of Value*, Indian Philosophy: A Collection of Readings 5, ed. Roy W. Perrett (Abingdon, UK: Routledge, 2011), 11–24.
7. R. P. Dangle, ed. and trans., *The Kauṭilīya Arthaśāstra Part II* (New Delhi, India: Motilal Banarsidass, 1986), 482.
8. Karen Armstrong, *Buddha* (London: Phoenix, 2000), 74.
9. For a deeper dive, seek out Padmasiri de Silva, "Theoretical Perspectives on Emotions in Early Buddhism," in *Emotions in Asian Thought: A Dialogue in Comparative Philosophy*, ed. Joel Marks and Roger T. Ames (Albany: State University of New York Press, 1995), 109–22.
10. This story is drawn from one of the works of the Buddhist Pali canon, *The Culasaccaka Sutta*, found in *The Middle Length Discourses of the Buddha: A Translation of the Majjhima Nikāya*, trans. Bhikkhu Ñānamoli and Bhikkhu Bodhi (Kandy, Sri Lanka: Buddhist Translation Society, 1995), 322–31.
11. *The Culasaccaka Sutta*, 322, 328.
12. *The Culasaccaka Sutta*, 323.
13. *The Culasaccaka Sutta*, 328, http://lirs.ru/lib/sutra/The_Middle_Length_Discourses (Majjhima_Nikaya),Nanamoli,Bodhi,1995.pdf.
14. *The Connected Length Discourses of the Buddha: A Translation of the Samyutta Nikāya*, trans. Bhikkhu Bodhi and Bhikkhu Ñānamoli (Kandy, Sri Lanka: Buddhist Translation Society, 2005), 421.
15. *Middle Length Discourses*, 121.

16. Ashoka, Major Rock Edict 13, in Romila Thapar, *Asoka and the Decline of the Mauryas* (Oxford, UK: Oxford University Press, 1961), 255–56.
17. Ashoka, Minor Pillar Edict 1, in Thapar, *Asoka and the Decline of the Mauryas*, 259.
18. Romila Thapar, "Aśoka and Buddhism as Reflected in the Aśokan Edicts," in *King Asoka and Buddhism: Historical and Literary Studies*, ed. Anuradha Seneviratna (Kandy, Sri Lanka: Buddhist Publication Society, 1995), 36.
19. According to *World Population Review*, https://worldpopulationreview.com/country-rankings/buddhist-countries.

Chapter Three: *The Pauline Passions*

1. Jan M. Bremmer, ed., *The Apocryphal Acts of Paul and Thecla*, Studies on the Apocryphal Acts of the Apostles 2 (Leuven, Belgium: Peeters Publishers, 1996), 38.
2. Acts 21:28. All the Bible verses in this book are from the Christian Standard Bible (CSB), with exceptions noted (Nashville, TN: B&H Publishing Group, 2020).
3. Acts 5:34.
4. Acts 9.
5. For the epilepsy idea, see D. Landsborough, "St Paul and Temporal Lobe Epilepsy," *Journal of Neurology, Neurosurgery, and Psychiatry* 40 (1987): 659–64; for the lightning suggestion, see John D. Bullock, "Was Saint Paul Struck Blind and Converted by Lightning?," *Survey of Ophthalmology* 39, no. 2 (September–October 1994): 151–60.
6. An oldie but goodie: Edward A. Wicher, "Ancient Jewish Views of the Messiah," *Journal of Religion* 34, no. 5 (November 1909): 317–25.
7. Leviticus 4:1–5:13.
8. Exodus 34:6–7. Jay P. Green, ed. and trans., *The Interlinear Bible: Hebrew-Greek-English*, 2nd ed. (Lafayette, IN: Sovereign Grace Publishers, 1997).
9. Deuteronomy 5:9–10.
10. She goes into way more detail in Valérie Curtis, *Don't Look, Don't Touch: The Science Behind Revulsion* (Oxford, UK: Oxford University Press, 2013).
11. Jaak Panksepp, "Criteria for Basic Emotions: Is DISGUST a Primary 'Emotion'?," *Cognition and Emotion* 21, no. 8 (2007): 1819–28.
12. *Shaqats:* Leviticus 11:10, 11:13, 11:43; Deuteronomy 7:26. *Sheqets:* Leviticus 11:10–13, 11:20, 11:44, 11:42; Isaiah 66:17; Ezekiel 8:10.
13. The Bibles used for cross-referencing were *The Latin and English Parallel Bible (Vulgate and KJV)* (Kirkland, WA: Latus ePublishing, 2011) and *The Interlinear Bible*, with variations in the Vulgate in parentheses below. For clarity, modern book names are retained.

    Toebah: Exodus 8:26; Leviticus 18:22, 18:26, 18:29, 20:13; Deuteronomy 7:25, 7:26, 13:14, 13:31, 17:1, 17:4, 18:9, 18:12, 20:18, 22:5, 23:18, 24:4, 25:16, 27:5, 32:16; 1 Kings 14:24; 2 Kings 21:11, 23:23; 2 Chronicles 33:2, 33:35, 36:8, 36:14; Ezra 9:1, 9:11, 9:14; Proverbs 3:32, 11:1, 11:20, 12:22, 15:8, 15:9, 15:26, 16:5, 16:12, 17:15, 20:10, 20:23, 21:27, 24:9, 29:27; Isaiah 1:13, 41:21; Jeremiah 2:7, 6:15, 7:10, 8:12, 16:18, 32:35, 44:2, 44:22; Ezekiel 5:9, 6:9, 6:11, 7:3, 7:4, 7:8, 8:6, 8:9, 8:13, 8:15, 8:17, 9:4, 9:16, 9:22, 9:36, 9:43, 9:50, 9:51, 18:21, 18:24, 20:4, 22:3, 22:11, 33:26, 33:29, 43:8; Malachi 2:11.

Taab: Job 9:31, 16:16, 19:19, 30:10; Psalms 5:6 (5:7), 14:1 (13:1), 53:1 (52:1, 53:2), 107:18 (106:18), 106:40 (105:40); Isaiah 65:4; Ezekiel 16:25; Amos 5:10; Micah 3:9.

14. Exodus 29:18 ("Then burn the whole ram on the altar; it is a burnt offering to the Lord. It is a pleasing aroma, a food offering to the Lord") and 29:25 ("Take them from their hands and burn them on the altar on top of the burnt offering, as a pleasing aroma before the Lord; it is a food offering to the Lord").
15. Acts 13:18.
16. If you really want to know more about Stoic logic, read Benson Mates, *Stoic Logic* (Socorro, NM: Advanced Reasoning Forum, 2014). Warning: it's not for the faint of heart!
17. Marcus Aurelius, *Meditations*, trans. and ed. Martin Hammond (London: Penguin Books, 2006), book 6, no. 13.
18. For more on the argument that the accepted order of events is wrong, see Christopher I. Beckwith, *Greek Buddha: Pyrrho's Encounter with Early Buddhism in Central Asia* (Princeton, NJ: Princeton University Press, 2015).
19. See Karen Armstrong, *The Great Transformation: The World in the Time of Buddha, Socrates, Confucius, and Jeremiah* (London: Atlantic Books, 2009), 367; Demetrios Th. Vassiliades, "Greeks and Buddhism: Historical Contacts in the Development of a Universal Religion," *Eastern Buddhist (New Series)* 36, no. 1–2 (2004): 134–83; Thomas C. McEvilley, *The Shape of Ancient Thought*, Comparative Studies in Greek and Indian Philosophies (New York: Allworth Press, 2006).
20. Diogenes Laërtius, *The Lives and Opinions of Eminent Philosophers*, trans. C. D. Yonge (London: G. Bell and Sons, 1915), book 9.
21. Acts 17:22.
22. Acts 17:24.
23. Acts 17:25.
24. Acts 17:26.
25. Acts 17:27.
26. Acts 17:28.
27. Acts 17:29.
28. Acts 17:30–31.
29. Acts 17:31.
30. Acts 17:32.
31. For the number of Christians globally, see the Pew Research Center study, https://www.pewresearch.org/fact-tank/2017/04/05/christians-remain-worlds -largest-religious-group-but-they-are-declining-in-europe/.

Chapter Four: *Crusader Love*

1. Helen Fisher, *Why We Love: The Nature and Chemistry of Romantic Love* (New York: Holt Paperbacks, 2005).
2. Kristyn R. Vitale Shreve, Lindsay R. Mehrkam, and Monique A. R. Udell, "Social Interaction, Food, Scent or Toys? A Formal Assessment of Domestic Pet and Shelter Cat (*Felis silvestris catus*) Preferences," *Behavioral Processes* 141, pt. 3 (August 2017): 322–28.

3. For a wonderful overview of belongingness, see Kelly-Ann Allen, *The Psychology of Belonging* (London: Routledge, 2020).

4. Robert C. Solomon, *About Love: Reinventing Romance for Our Times* (New York: Simon & Schuster, 1988); Mark Fisher, *Personal Love* (London: Duckworth, 1990).

5. Gabriele Taylor, "Love," *Proceedings of the Aristotelian Society (New Series)* 76 (1975–1976): 147–64; Richard White, *Love's Philosophy* (Oxford, UK: Rowman & Littlefield, 2001).

6. J. David Velleman, "Love as a Moral Emotion," *Ethics* 109, no. 2 (1999): 338–74.

7. Saint Augustine, *Confessions*, trans. Henry Chadwick, Oxford World's Classics (Oxford, UK: Oxford University Press, 2009).

8. Saint Augustine, *Confessions*, 96–97.

9. Saint Augustine, *Confessions*, 97.

10. Saint Augustine, *Confessions*, 97.

11. Saint Augustine, *Confessions*, 98–99.

12. Romans 13:13.

13. Genesis 27.

14. Mark 12:30–31.

15. Helmut David Baer, "The Fruit of Charity: Using the Neighbor in *De doctrina christiana*," *Journal of Religious Ethics* 24, no. 1 (Spring 1996): 47–64.

16. Fulcher of Chartres, "The Speech of Urban II at the Council of Clermont, 1095," in *A Source Book for Mediæval History: Selected Documents Illustrating the History of Europe in the Middle Age*, ed. Oliver J. Thatcher and Edgar Holmes McNeal, trans. Oliver J. Thatcher (New York: Charles Scribner's Sons, 1905), 513–17.

17. Fulcher of Chartres, "The Speech of Urban II."

18. Translation found in August C. Krey, *The First Crusade: The Accounts of Eyewitnesses and Participants* (Princeton, NJ: Princeton University Press, 1921), 19.

19. Dana C. Munro, ed., "Urban and the Crusaders," in *Translations and Reprints from the Original Sources of European History* (Philadelphia: University of Pennsylvania Press, 1895), 1:5–8.

20. Krey, *The First Crusade*, 18.

21. Imad ad-Din al-Isfahani, in *Arab Historians of the Crusades*, ed. Francesco Gabrieli, trans. E. J. Costello (Abingdon, UK: Routledge, 2010), 88.

22. Saint Augustine, *The Works of Aurelius Augustine, Bishop of Hippo: A New Translation*, ed. Rev. Marcus Dods, M.A., vol. 1, *The City of God*, trans. Rev. Marcus Dods, M.A. (Edinburgh: T. & T. Clark, 1871), 33.

23. Krey, *The First Crusade*, 42.

24. Krey, *The First Crusade*, 42.

Chapter Five: *What the Ottomans Feared*

1. Vani Mehmed Efendi, "'Ara'is al-Kur'an Wa Nafa'is al-Furkan [Ornaments of the Quran and the Valuables of the Testament]" (Yeni Cami 100, Istanbul, 1680), para. 543a, Suleymaniye Library; as translated in Mark David Baer, *Honored by the Glory of Islam: Conversion and Conquest in Ottoman Europe* (Oxford, UK: Oxford University Press, 2008), 207.

2. Thierry Steimer, "The Biology of Fear- and Anxiety-Related Behaviours," *Dialogues in Clinical Neuroscience* 4, no. 3 (2002): 231–49.

3. N. J. Dawood, trans., *The Koran* (London: Penguin Classics, 1978), 418.

4. Ibn Ishaq, "The Hadith," in *Islam*, ed. John Alden Williams (New York: George Braziller, 1962), 61.

5. Abdur-Rahman bin Saib, *hadith* 1337, in "The Chapters of Establishing the Prayer and the Sunnah Regarding Them," chapter 7 of *Sunan Ibn Majah*, Ahadith.co.uk, https://ahadith.co.uk/chapter.php?cid=158&page=54&rows=10, accessed August 20, 2020.

6. For a fantastic overview survey of Koranic emotions, see Karen Bauer, "Emotion in the Qur'an: An Overview" [Edited Version], *Journal of Qur'anic Studies*, 19, no. 2 (2017): 1–31.

7. Dawood, *The Koran*, 30:38.

8. Dawood, *The Koran*, 3:174.

9. الخوف من الله

10. Dawood, *The Koran*, 103:1–3.

11. Dawood, *The Koran*, 384.

12. Bauer, "Emotion in the Qur'an," 18.

13. Dawood, *The Koran*, 22:46.

14. Regarding Galen not mentioning humoral causes of the passions, see Galen, *On the Passions and Errors of the Soul*, trans. Paul W. Harkins (Columbus: Ohio State University Press, 1963).

15. A keen reader might have noticed that Galen split the body into the same three areas that Plato associated with the three-part soul—lower body, heart/chest, and brain. This wasn't accidental: Galen was a bit of a Platonist.

16. This edited version is a little dated in some of its commentaries, but the translation of Ibn Sina's text is apparently quite good: Ibn Sina (Avicenna), *The Canon of Medicine of Avicenna*, trans. Oscar Cameron Gruner (New York: AMS Press, 1973), 285, 321.

17. Mevlâna Mehmet Neşri, *Gihānnümā [Cihannüma] Die Altosmanische Chronik Des Mevlānā Meḥemmed Neschrī [Mevlâna Mehmet Neşri]*, ed. Franz Taeschner (Wiesbaden, Germany: Harrassowitz Verlag, 1951), 194, as translated in Halil Inalcik, *The Ottoman Empire: The Classical Age 1300–1600* (London: Weidenfeld & Nicolson, 2013), loc. 5109, Kindle.

18. Nil Tekgül, "A Gate to the Emotional World of Pre-Modern Ottoman Society: An Attempt to Write Ottoman History from 'the Inside Out'" (PhD diss., Bilkent University, 2016), 177, http://repository.bilkent.edu.tr/handle/11693/29154, accessed February 20, 2020.

19. Dawood, *The Koran*, 23:51.

20. Tekgül, "A Gate to the Emotional World," 84–87.

Chapter Six: *Abominable Witch Crazes*

1. I'd be remiss if I didn't mention the following book. It has a different take on witches and emotions from mine, but it's nevertheless superb: Charlotte-Rose Millar, *Witchcraft, the Devil, and Emotions in Early Modern England* (Abingdon, UK: Routledge, 2017).

2. William Rowley, Thomas Dekker, and John Ford, *The Witch of Edmonton* (London: J. Cottrel for Edward Blackmore, 1658).

3. Malcolm Gaskill claims that the number is between forty-five and fifty thousand: see Malcolm Gaskill, *Witchcraft: A Very Short Introduction* (Oxford, UK: Oxford University Press, 2010), 76. Brian Levack claims sixty thousand: see Brian P. Levack, *The Witch-Hunt in Early Modern Europe* (London: Longman, 2013), 22. Anne Llewellyn Barstow claims that, all records taken together, it's at least one hundred thousand: see Anne Llewellyn Barstow, *Witchcraze* (London: Bravo, 1995). I could go on. Almost any book on the subject will give you a different figure. The important point is, it was a lot.

4. In Latin, *amor* and *odium*, *spes* and *desperatio*, *audacia* and *timor*, *gaudium* and *tristitia*. See Thomas Aquinas, *The Emotions (Ia2æ. 22–30)*, vol. 19 of *Summa Theologiae*, ed. Eric D'Arcy (Cambridge, UK: Blackfriars, 2006), XVI, Q. 23.

5. Frank Tallett, *War and Society in Early-Modern Europe, 1495–1715* (Abingdon, UK: Routledge, 1992), 13.

6. The classic text on the Little Ice Age is Emmanuel Le Roy Ladurie, *Times of Feast, Times of Famine: A History of Climate Since the Year 1000* (New York: Doubleday, 1971); for something more up to date, read Brian Fagan, *The Little Ice Age: How Climate Made History, 1300–1850*, rev. ed. (New York: Basic Books, 2019).

7. The English sweat was first described in John Caius, *A Boke or Counseill Against the Disease Commonly Called the Sweate, or Sweatyng Sicknesse*, published in 1552.

8. For more on this subject, see Deborah Hayden, *Pox: Genius, Madness, and the Mysteries of Syphilis* (New York: Basic Books, 2003); Mircea Tampa et al., "Brief History of Syphilis," *Journal of Medical Life* 7, no. 1 (2014): 4–10.

9. William Shakespeare, *The Rape of Lucrece*, http://shakespeare.mit.edu/Poetry/RapeOfLucrece.html.

10. Carol Nemeroff and Paul Rozin, "The Contagion Concept in Adult Thinking in the United States: Transmission of Germs and of Interpersonal Influence," *Ethos* 22, no. 2 (June 1994): 158–86.

11. Bruce M. Hood, *SuperSense: Why We Believe in the Unbelievable* (San Francisco: HarperOne, 2009), 139, 170.

12. Hood, *SuperSense*, 215–16.

13. Hood, *SuperSense*, 139, 170.

14. Robert Ian Moore, *The Formation of a Persecuting Society: Power and Deviance in Western Europe, 950–1250* (Oxford, UK: Basil Blackwell, 1987), 64.

15. Charles Zika, *The Appearance of Witchcraft: Print and Visual Culture in Sixteenth-Century Europe* (Abingdon, UK: Routledge, 2007), 80–81.

16. Francesco Maria Guazzo, *Compendium Maleficarum: The Montague Summers Edition* (Mineola, NY: Dover Publications, 1988), 11, 35.

17. Heinrich Kramer and Jacob Sprenger, *The Malleus Maleficarum*, ed. and trans. P. G. Maxwell-Stuart (Manchester, UK: Manchester University Press, 2007), 184, 231.
18. Luana Colloca and Arthur J. Barsky, "Placebo and Nocebo Effects," *New England Journal of Medicine* 382, no. 6 (February 6, 2020): 554–61.

Chapter Seven: *A Desire for Sweet Freedom*

1. John Locke, *Two Treatises of Government* (London: Whitmore & Fenn, 1821), 189, 191, 199, 209, https://books.google.com/books?id=K5UIAAAAQAAJ&printsec =frontcover&source=gbs_ge_summary_r&cad=0#v=onepage&q&f=false.
2. David Hume, *An Enquiry Concerning the Principles of Morals* (Indianapolis: Hackett Publishing, 1983), 3:12.
3. John K. Alexander, *Samuel Adams: America's Revolutionary Politician* (Oxford, UK: Rowman & Littlefield, 2002), 125; Ray Raphael, *A People's History of the American Revolution: How Common People Shaped the Fight for Independence* (New York: New Press, 2001), 18.
4. Harry G. Frankfurt, *On Bullshit* (Princeton, NJ: Princeton University Press, 2005); Harry G. Frankfurt, "Freedom of the Will and the Concept of a Person," *Journal of Philosophy* 68, no. 1 (January 14, 1971): 5–20.
5. Aristotle, "Sense and Sensibilia," trans. J. I. Beare, in *The Complete Works of Aristotle*, ed. Jonathan Barnes (Princeton, NJ: Princeton University Press, 1984), 1:693–713, 436b15–446a20.
6. For a good take on our strange relationship with our orifices, see William Ian Miller, *The Anatomy of Disgust* (Cambridge, MA: Harvard University Press, 1997), 89–98.
7. Anselm of Canterbury, "Liber Anselmi Archiepiscopi de Humanis Moribus," in *Memorials of St Anselm*, ed. Richard William Southern and F. S. Schmitt (Oxford, UK: Oxford University Press, 1969), 47–50.
8. Niall Atkinson, "The Social Life of the Senses: Architecture, Food, and Manners," in *A Cultural History of the Senses*, vol. 3, *In the Renaissance*, ed. Herman Roodenburg (London: Bloomsbury, 2014), 33.
9. Especially popular was Cicero, *De Officiis*, trans. Walter Miller (Cambridge, MA: Harvard University Press, 1913).
10. Maestro Martino, *Libro de arte coquinaria*, www.loc.gov/item/2014660856/.
11. I translated this from the original early modern English text: "Ther be many cristen bothe clerkes and layemen whyche lyl know god by fayth ne by scrupture by cause they haue the taste disordynate by synne they may not wel sauoure hym." See Gui de Roye, *Thus Endeth the Doctrinal of Sapyence*, trans. Wyllyam Caxton (Cologne: Wyllyam Caxton, 1496), fol. 59r, https://tinyurl.com/uv56 xekm.
12. A great overview of luxury in the period can be found in Linda Levy Peck, *Consuming Splendor: Society and Culture in Seventeenth-Century England* (Cambridge, UK: Cambridge University Press, 2005).
13. Bernard Mandeville, *The Fable of the Bees* (London: T. Ostell, 1806), 66–73.
14. Anthony Ashley Cooper, 3rd Earl of Shaftesbury, *Characteristicks of Men, Manners, Opinions, Times* (Carmel, IN: Liberty Fund, 2001), 2:239.

15. Francis Hutcheson, *An Inquiry into the Original of Our Ideas of Beauty and Virtue* (London: J. Darby, 1726), 73.
16. For a good example, see Samuel Clarke, *A Demonstration of the Being and Attributes of God: And Other Writings*, ed. Ezio Vailati, Cambridge Texts in the History of Philosophy (Cambridge, UK: Cambridge University Press, 1998).
17. The best book on this subject is undoubtedly Thomas Dixon, *From Passions to Emotions: The Creation of a Secular Psychological Category* (Cambridge, UK: Cambridge University Press, 2003).
18. Adam Smith, *The Theory of Moral Sentiments*, ed. Knud Haakonssen, Cambridge Texts in the History of Philosophy (Cambridge, UK: Cambridge University Press, 2002).
19. Smith, *The Theory of Moral Sentiments*, 209–10, 218–20, 227–34.

Chapter Eight: *Becoming Emotional*

1. René Descartes, *The Passions of the Soule*, anonymous translator (London: 1650), answer to second letter, B3r–B3v. Some translators have rendered the original French word *physicien* as either "physicist" or "natural philosopher." This seems odd, given that the text is overwhelmingly medical. See https://quod.lib.umich.edu/cgi/t/text/text-idx?c=eebo2;idno=A81352.0001.001.
2. Descartes, *Passions*, article 46.
3. Thomas Hobbes, *Leviathan*, ed. Noel Malcolm, Clarendon Edition of the Works of Thomas Hobbes (Oxford, UK: Oxford University Press, 2012), 2:78.
4. Hobbes, *Leviathan*, 2:84.
5. Hobbes, *Leviathan*, 2:84.
6. Hobbes, *Leviathan*, 2:84.
7. Hobbes, *Leviathan*, 2:84.
8. David Hume, *A Treatise of Human Nature*, 2.3.3.4, https://davidhume.org/texts/t/2/3/3#4.
9. Thomas Brown, *A Treatise on the Philosophy of the Human Mind*, ed. Levi Hodge (Cambridge, UK: Hilliard and Brown, 1827), 1:103.
10. William James, "What Is an Emotion?," *Mind* 9, no. 34 (April 1884): 190.
11. Paul R. Kleinginna Jr. and Anne M. Kleinginna, "A Categorized List of Emotion Definitions, with Suggestions for a Consensual Definition," *Motivation and Emotion* 5, no. 4 (1981): 345–79.

Chapter Nine: *A Cherry-Blossomed Shame*

1. Robert Louis Stevenson, "Yoshida-Torajiro," in *The Works of Robert Louis Stevenson*, vol. 2, *Miscellanies: Familiar Studies of Men and Books* (Edinburgh: T. and A. Constable, 1895), 165.
2. Yoshida Shōin, "Komo Yowa," translated in Eiko Ikegami, "Shame and the Samurai: Institutions, Trustworthiness, and Autonomy in the Elite Honor Culture," *Social Research* 70, no. 4 (Winter 2003): 1354.
3. Gershen Kaufman, *The Psychology of Shame: Theory and Treatment of Shame-Based Syndromes*, 2nd ed. (New York: Springer, 1989); Kelly McGonigal, *The Upside of Stress: Why Stress Is Good for You (and How to Get Good at It)* (London: Vermilion,

2015); Paul Gilbert, *The Compassionate Mind* (London: Constable, 2010); Joseph E. LeDoux, "Feelings: What Are They & How Does the Brain Make Them?," *Dædalus* 144, no. 1 (January 2015): 105.

4. E. Tory Higgins et al., "Ideal Versus Ought Predilections for Approach and Avoidance Distinct Self-Regulatory Systems," *Journal of Personality and Social Psychology* 66, no. 2 (February 1994): 276–86.

5. Zisi, "Zhong Yong," trans. James Legge, Chinese Text Project, https://ctext.org /liji/zhong-yong, accessed November 23, 2020; Donald Sturgeon, "Chinese Text Project: A Dynamic Digital Library of Premodern Chinese," *Digital Scholarship in the Humanities*, August 29, 2019.

6. William E. Deal and Brian Ruppert, *A Cultural History of Japanese Buddhism*, Wiley Blackwell Guides to Buddhism (Oxford, UK: John Wiley & Sons, 2015), 172.

7. Fumiyoshi Mizukami, "Tenkai no isan: Tenkaihan issaikyō mokukatsuji," in *Minshū bukkyō no teichaku*, ed. Sueki Fumihiko (Tokyo: Kōsei, 2010), 125.

8. Philip Kapleau, *The Three Pillars of Zen: Teaching, Practice, and Enlightenment* (New York: Anchor, 1989), 85.

9. Jakuren, *Shinkokinshū* 4:361, trans. Thomas McAuley, Waka Poetry, http://www .wakapoetry.net/skks-iv-361/, accessed November 20, 2020.

10. Royall Tyler, ed. and trans., *Japanese Nō Dramas* (London: Penguin Classics, 1992), 72–73.

11. Gary L. Ebersole, "Japanese Religions," in *The Oxford Handbook of Religion and Emotion*, ed. John Corrigan (Oxford, UK: Oxford University Press, 2008), 86.

12. Gian Marco Farese, "The Cultural Semantics of the Japanese Emotion Terms 'Haji' and 'Hazukashii,'" *New Voices in Japanese Studies* 8 (July 2016): 32–54.

13. Yoshida, "Komo Yowa," 1353.

Chapter Ten: *The Rage of an African Queen*

1. Edwin W. Smith, *The Golden Stool: Some Aspects of the Conflict of Cultures in Modern Africa* (London: Holborn Publishing House, 1926), 5.

2. The speech came from field notes taken by Agnes Aidoo in 1970 and related by an eyewitness, Opanin Kwabena Baako. See Agnes Akosua Aidoo, "Asante Queen Mothers in Government and Politics in the Nineteenth Century," *Journal of the Historical Society of Nigeria* 9, no. 1 (December 1977): 12.

3. R. J. R. Blair, "Considering Anger from a Cognitive Neuroscience Perspective," *Wiley Interdisciplinary Reviews: Cognitive Science* 3, no. 1 (January–February 2012): 65–74.

4. Kwame Gyekye, *An Essay on African Philosophical Thought: The Akan Conceptual Scheme*, rev. ed. (Philadelphia: Temple University Press, 1995), 85–88.

5. Gyekye, *An Essay on African Philosophical Thought*, 88–94.

6. Gyekye, *An Essay on African Philosophical Thought*, 95–96.

7. Gyekye, *An Essay on African Philosophical Thought*, 95.

8. Gyekye, *An Essay on African Philosophical Thought*, 100.

9. Peter Sarpong, *Ghana in Retrospect: Some Aspects of Ghanaian Culture* (Accra: Ghana Publishing Corporation, 1974), 37; Meyer Fortes, *Kinship and the Social*

Order: The Legacy of Lewis Henry Morgan (Chicago: University of Chicago Press, 1969), 199n14; Gyekye, *An Essay on African Philosophical Thought*, 94.

10. Much of the linguistic data comes from this fantastic paper on the subject: Vivian Afi Dzokoto and Sumie Okazaki, "Happiness in the Eye and the Heart: Somatic Referencing in West African Emotion Lexica," *Journal of Black Psychology* 32, no. 2 (2006): 117–140.

11. This paper is a wonderful in-depth analysis of Akan proverbs: Vivian Afi Dzokoto et al., "Emotion Norms, Display Rules, and Regulation in the Akan Society of Ghana: An Exploration Using Proverbs," *Frontiers in Psychology* 9 (2018), https://www.frontiersin.org/article/10.3389/fpsyg.2018.01916.

12. Andy Clark, *Being There: Putting Brain, Body, and World Together Again* (Cambridge, MA: MIT Press, 1997), xii.

13. See Glenn Adams, "The Cultural Grounding of Personal Relationship: Enemyship in North American and West African Worlds," *Journal of Personality and Social Psychology* 88, no. 6 (June 2005): 948–68.

14. Adams, "The Cultural Grounding."

15. Gladys Nyarko Ansah, "Emotion Language in Akan: The Case of Anger," in *Encoding Emotions in African Languages*, ed. Gian Claudio Batic (Munich: LINCOM GmbH, 2011), 131.

16. Ansah, "Emotion Language," 134.

17. Ansah, "Emotion Language," 131.

18. See T. C. McCaskie, "The Life and Afterlife of Yaa Asantewaa," *Africa: Journal of the International African Institute* 77, no. 2 (2007): 170.

Chapter Eleven: *Shell Shocks*

1. Based on an undated and untitled typescript by W. D. Esplin, Public Record Office, Kew, PRO PIN15/2502.

2. G. Elliot Smith and T. H. Pear, *Shell Shock and Its Lessons* (Manchester, UK: Manchester University Press, 1918), 12–13.

3. A good overview of CSR is found in Zahava Solomon, *Combat Stress Reaction: The Enduring Toll of War*, Springer Series on Stress and Coping (New York: Springer, 2013).

4. Cowardice was the underlying diagnosis of E. D. Adrian and L. R. Yealland, "The Treatment of Some Common War Neuroses," *Lancet* 189, no. 4893 (June 9, 1917): 867–72.

5. See, for example, Thomas Dixon, *Weeping Britannia: Portrait of a Nation in Tears* (Oxford, UK: Oxford University Press, 2015), 201–2; Tracey Loughran, *Shell-Shock and Medical Culture in First World War Britain*, Studies in the Social and Cultural History of Modern Warfare 48, reprint ed. (Cambridge, UK: Cambridge University Press, 2020), 115.

6. Rudyard Kipling, "If—" (1943), Poetry Foundation, https://www.poetryfoundation.org/poems/46473/if---, accessed August 19, 2020.

7. Sigmund Freud, *Letters of Sigmund Freud*, ed. Ernst L. Freud, trans. Tania and James Stern (Mineola, NY: Dover Publications, 1992), 175.

8. Jean-Martin Charcot, *Oeuvres complètes de J. M. Charcot: Leçons sur les maladies du système nerveux, faites à la Salpêtrière* (Paris: Bureaux du Progrès Médical / A. Delahaye & E. Lacrosnier, 1887), 3:436–62 (lecture 26).

9. Jean-Martin Charcot, *Leçons sur les maladies du système nerveux*, 12.

10. Sigmund Freud, "The Unconscious," in *The Standard Edition of the Complete Psychological Works of Sigmund Freud*, ed. and trans. James Strachey, vol. 14 (1914–1916), *On the History of the Psycho-Analytic Movement, Papers on Metapsychology and Other Works* (London: Hogarth Press, 1957), 159–216.

11. R. H. Cole, *Mental Diseases: A Text-Book of Psychiatry for Medical Students and Practitioners* (London: University of London Press, 1913), 47.

12. Cole, *Mental Diseases*, 47–48.

13. Cole, *Mental Diseases*, 48.

14. Cole, *Mental Diseases*, 49.

15. Cole, *Mental Diseases*, 51.

16. Cole, *Mental Diseases*, 52.

17. T. C. Shaw, *Ex Cathedra: Essays on Insanity* (London: Adlard and Sons, 1904), 110.

18. Siegfried Sassoon, "Declaration Against the War," in Robert Giddings, *The War Poets: The Lives and Writings of the 1914–18 War Poets* (London: Bloomsbury, 1990), 111.

19. Siegfried Sassoon, *The War Poems of Siegfried Sassoon* (London: William Heinemann, 1919), 43–44.

20. Sigmund Freud, "Five Lectures on Psychoanalysis," in *The Standard Edition of the Complete Psychological Works of Sigmund Freud*, ed. and trans. James Strachey, vol. 11 (1910), *Five Lectures on Psycho-Analysis, Leonardo da Vinci, and Other Works* (London: Hogarth Press, 1957), 49.

21. Rebecca West, *The Return of the Soldier* (London: Virago Modern Classics, 2010).

Chapter Twelve: *The Dragon's Humiliation*

1. As described in Yung-fa Chen, *Making Revolution: The Communist Movement in Eastern and Central China, 1937–1945* (Berkeley: University of California Press, 1986), 186–87.

2. As a side note, there are some historians, such as Russell Kirkland, who don't think Chang-tzŭ ever existed and was the invention of a man named Kuo Hsiang, writing six hundred years later. But we do have a much earlier biography of him, written by a man named Sima Qian just one hundred years after Chang-tzŭ died. It's impossible to know for sure. What we do know is that the wisdom contained in the book he supposedly wrote, the *Zhuangzi*, is an important text in the history of Daoism. So I'm going to treat him as if he existed. See Russell Kirkland, *Taoism: The Enduring Tradition* (Abingdon, UK: Routledge, 2004), 33–34.

3. Chang-tzŭ, *The Inner Chapters*, trans. A. C. Graham (Indianapolis: Hackett Publishing, 2001), 120–21.

4. Confucius, *The Analects*, ed. and trans. Raymond Dawson, Oxford World's Classics (Oxford, UK: Oxford University Press, 2008), 17:21. By the way, if these Chinese-language terms seem unfamiliar, then I will assume you are doing as I

suggested and are first reading the parts of this book that most interest you. In that case, all will become clear if you read chapter 9.

5. Chang-tzǔ, *The Inner Chapters*, 123–24.

6. Chang-tzǔ, *The Inner Chapters*, 211.

7. Laozi, *Daodejing: The New, Highly Readable Translation of the Life-Changing Ancient Scripture Formerly Known as the Tao Te Ching*, ed. and trans. Hans-Georg Moeller (Chicago: Open Court Publishing, 2007), 51.

8. Alan Watts, *Tao: The Watercourse Way* (New York: Pantheon Books, 1975), 45–46.

9. See chapter 1 of the Liuzi (Master Liu), in *Baizi Quanshu (A Complete Collection of Works by the One Hundred Masters)* (Shanghai: Zhejiang Renmin Chubanshe, 1991), 6:1, as translated in Heiner Fruehauf, "All Disease Comes from the Heart: The Pivotal Role of the Emotions in Classical Chinese Medicine," *Journal of Chinese Medicine* (2006): 2.

10. For a definition, see Bob Flaws and James Lake, MD, *Chinese Medical Psychiatry: A Textbook and Clinical Manual* (Portland, OR: Blue Poppy Enterprises, 2001).

11. Lin Zexu, letter to Queen Victoria, 1839, trans. Mitsuko Iriye and Jerome S. Arkenberg, in *Modern Asia and Africa*, ed. William H. McNeill and Mitsuko Iriye, Readings in World History 9 (Oxford, UK: Oxford University Press, 1971), 111–18.

12. Elizabeth J. Perry, "Moving the Masses: Emotion Work in the Chinese Revolution," *Mobilization* 7, no. 2 (2002): 112.

13. Edgar Snow, *Red Star over China: The Classic Account of the Birth of Chinese Communism*, rev. ed. (London: Grove Press, 2007), loc. 1900, Kindle.

14. Perry, "Moving the Masses," 113.

15. Mao Zedong, "The Chinese People Have Stood Up!," in *Selected Works of Mao Tse-tung*, https://www.marxists.org/reference/archive/mao/selected-works/volume-5/mswv5_01.htm.

16. Liang Heng and Judith Shapiro, *Son of the Revolution* (New York: Vintage, 1984), 77–79; also in Perry, "Moving the Masses," 122.

17. Perry, "Moving the Masses," 122.

Chapter Thirteen: *Love and the Mother(land)*

1. "John F. Kennedy Moon Speech—Rice Stadium: September 12, 1962," NASA Space Educators' Handbook, https://er.jsc.nasa.gov/seh/ricetalk.htm, accessed September 5, 2019.

2. Walter Rugaber, "Nixon Makes 'Most Historic Telephone Call Ever,'" *New York Times*, July 21, 1969.

3. John Lear, "Hiroshima, U.S.A.: Can Anything Be Done About It?," *Collier's*, August 5, 1950, 12, https://www.unz.com/print/Colliers-1950aug05-00011.

4. Mick Jackson, dir., *Threads* (BBC, 1984).

5. Howard S. Liddell, "Conditioning and Emotions," *Scientific American* 190, no. 1 (January 1954): 48.

6. Semir Zeki and John Paul Romaya, "Neural Correlates of Hate," *PLoS One* 3, no. 10 (2008): e3556, https://doi.org/10.1371/journal.pone.0003556.

7. For one of many discussions by Freud on *ambivalenz*, see Sigmund Freud, *Totem und Tabu* (Vienna: Vienna University Press, 2013), 77–123; for a good version in English, see Sigmund Freud, *Totem and Taboo* (Abingdon, UK: Routledge, 2012), 21–86.

8. Please take with a grain of salt. See Zeki and Romaya, "Neural Correlates of Hate"; Andreas Bartels and Semir Zeki, "The Neural Basis of Romantic Love," *NeuroReport* 11, no. 17 (November 2000): 3829–33; Wang Jin, Yanhui Xiang, and Mo Lei, "The Deeper the Love, the Deeper the Hate," *Frontiers in Psychology* 8, no. 1940 (December 2017), https://doi.org/10.3389/fpsyg.2017.01940, accessed June 3, 2019.

9. Kathryn J. Lively and David R. Heise, "Sociological Realms of Emotional Experience," *American Journal of Sociology* 109, no. 5 (March 2004): 1109–36; Elizabeth Williamson, "The Magic of Multiple Emotions: An Examination of Shifts in Emotional Intensity During the Reclaiming Movement's Recruiting/Training Events and Event Reattendance," *Sociological Forum* 26, no. 1 (March 2011): 45–70.

10. James M. Jasper, "Emotions and Social Movements: Twenty Years of Theory and Research," *Annual Review of Sociology* 37, no. 1 (August 2011): 285–303.

11. William Shakespeare, *Romeo and Juliet*, in *The Complete Works of William Shakespeare* (London: Wordsworth Editions, 2007), 256.

12. For a deeper study, see Guy Oakes, *The Imaginary War: Civil Defense and American Cold War Culture* (Oxford, UK: Oxford University Press, 1994), 47.

13. Frederick Peterson, "Panic—the Ultimate Weapon?," *Collier's*, August 21, 1953, 109.

14. Kelly A. Singleton, "The Feeling American: Emotion Management and the Standardization of Democracy in Cold War Literature and Film" (PhD diss., University of Maryland, 2017), https://drum.lib.umd.edu/bitstream/handle/1903/19372/Singleton_umd_0117E_17874.pdf.

15. Daniel Bell, "The End of Ideology in the West," in *The New York Intellectuals Reader*, ed. Neil Jumonville (New York: Routledge, 2007), 199.

16. Margaret Mead, *And Keep Your Powder Dry: An Anthropologist Looks at America* (Oxford, UK: Berghahn Books, 2005), 41.

17. See also Margaret Mead, *Soviet Attitudes Toward Authority* (New York: McGraw-Hill, 1951).

18. This included Ruth Benedict, *The Chrysanthemum and the Sword: Patterns of Japanese Culture* (New York: Houghton Mifflin, 1946); Geoffrey Gorer and John Rickman, *The People of Great Russia: A Psychological Study* (London: Cresset Press, 1949); Theodor W. Adorno et al., *The Authoritarian Personality* (New York: Harper and Brothers, 1950).

19. Lawrence K. Frank and Mary Frank, *How to Be a Woman* (Whitefish, MT: Literary Licensing, 2011).

20. John Bowlby, *Maternal Care and Mental Health: A Report Prepared on Behalf of the World Health Organization as a Contribution to the United Nations Programme for the Welfare of Homeless Children*, 2nd ed. (Geneva: World Health Organization, 1952), 12.

21. Bowlby, *Maternal Care and Mental Health*, 67.
22. Y. V. Popov and A. E. Lichko, "A Somber Page in the History of the All-Union Psychiatric Association," *Bekhterev Review of Psychiatry and Medical Psychology* 3 (1991): 116– 120.
23. I. Vainstajn, review of A. Zalkind's book *Ocerk kultury revoljucionnogo vremeni* (Essay of a culture of a revolutionary time), *Pod znamenem marksizma* (Under the banner of Marxism) (1924): 4–5, 297–300. For an English-language source, see Levy Rahmani, "Social Psychology in the Soviet Union," *Studies in Soviet Thought* 13, nos. 3–4 (September–December 1973): 221.
24. A. V. Zalkind, *Ocerk kultury revoljucionnogo vremeni* (Moscow: Rabotnik Prosvescenija, 1924).
25. A. V. Zalkind, "Psikhonevrologicheskie Mauki i Sotsialisticheskoe Stroitelstvo," *Pedologia* 3 (1930): 309–22; Alexander Etkind, "Psychological Culture," in *Russian Culture at the Crossroads: Paradoxes of Postcommunist Consciousness*, ed. Dmitri N. Shalin (Boulder, CO: Westview Press, 1996). Translations found in Dmitri N. Shalin, "Soviet Civilization and Its Emotional Discontents," *International Journal of Sociology and Social Policy* 16, nos. 9–10 (October 1996): 26.
26. An interesting read on this difference as it still exists is Maria A. Gartstein et al., "A Cross-Cultural Study of Infant Temperament: Predicting Preschool Effortful Control in the United States of America and Russia," *European Journal of Developmental Psychology* 6, no. 3 (May 2009): 337–64.

Chapter Fourteen: *The Great Emotions Face-Off*

1. This actually happened to my wife.
2. See "In Search of Universals in Human Emotion with Dr. Paul Ekman," Exploratorium, 2008, https://www.exploratorium.edu/video/search-universals-human -emotion-dr-paul-ekman, accessed October 19, 2018.
3. Margaret Mead, *Coming of Age in Samoa: A Psychological Study of Primitive Youth for Western Civilisation* (New York: William Morrow, 1928).
4. Paul Ekman and Wallace V. Friesen, "Constants Across Cultures in the Face and Emotion," *Journal of Personality and Social Psychology* 17, no. 2 (1971): 124–29; Paul Ekman, E. Richard Sorenson, and Wallace V. Friesen, "Pan-Cultural Elements in Facial Displays of Emotion," *Science* 164, no. 3875 (April 1969): 86–88.
5. "In Search of Universals in Human Emotion with Dr. Paul Ekman."
6. James Russell, "Language, Emotion, and Facial Expression" (lecture given at the fifteenth Kráków Medical Conference [The Emotional Brain: From the Humanities to Neuroscience, and Back Again], Copernicus Center for Interdisciplinary Studies, Kráków, Poland, May 20, 2011), https://youtu.be/oS1ZtvrgDLM, accessed November 26, 2011; *The Psychology of Facial Expression*, ed. James A. Russell and José Miguel Fernández-Dols, Studies in Emotion and Social Interaction (Cambridge, UK: Cambridge University Press, 1997); Sherri C. Widen and James A. Russell, "Children's Scripts for Social Emotions: Causes and Consequences Are More Central Than Are Facial Expressions," *British Journal of Developmental Psychology* 28 (September 2010): 565–81; James A. Russell and Beverley Fehr, "Relativity in the Perception of Emotion in Facial Expressions,"

Journal of Experimental Psychology: General 116, no. 3 (September 1987): 223–37; James M. Carroll and James Russell, "Do Facial Expressions Signal Specific Emotions? Judging Emotion from the Face in Context," *Journal of Personality and Social Psychology* 70, no. 2 (February 1996): 205–18.

7. Lisa Feldman Barrett et al., "Emotional Expressions Reconsidered: Challenges to Inferring Emotion from Human Facial Movements," *Psychological Science in the Public Interest* 20, no. 1 (2019): 1–68; see also Lisa Feldman Barrett, *How Emotions Are Made: The Secret Life of the Brain* (New York: Houghton Mifflin Harcourt, 2017), 4–12.

8. Stanley Schachter and Jerome Singer, "Cognitive, Social, and Physiological Determinants of Emotional State," *Psychological Review* 69, no. 5 (1962): 379–99.

9. It has been argued that postmodern art began much earlier in Europe, possibly as far back as 1915.

10. Catherine A. Lutz, *Unnatural Emotions: Everyday Sentiments on a Micronesian Atoll and Their Challenge to Western Theory* (Chicago: University of Chicago Press, 1988), 44–45.

11. Lutz, *Unnatural Emotions*, 16.

12. Lutz, *Unnatural Emotions*, 126.

13. Lutz, *Unnatural Emotions*, 131.

14. William Ian Miller, *The Anatomy of Disgust* (Cambridge, MA: Harvard University Press, 1997), 247; George Orwell, *The Road to Wigan Pier* (New Delhi: Delhi Open Books, 2019), 125, 131.

15. Jonathan Haidt, "The Disgust Scale Home Page," New York University Stern School of Business, 2012, http://people.stern.nyu.edu/jhaidt/disgustscale.html, accessed August 1, 2020.

16. Simone Schnall et al., "Disgust as Embodied Moral Judgment," *Personality and Social Psychology Bulletin* 34, no. 8 (May 2008): 1096–1109; Jonathan Haidt, "The Moral Emotions," in *Handbook of Affective Sciences*, ed. Richard J. Davidson, Klaus R. Scherer, and H. Hill Goldsmith, Series in Affective Science (Oxford, UK: Oxford University Press, 2003), 852–70.

17. Jonathan Haidt, "The Disgust Scale, Version 1," New York University Stern School of Business, http://people.stern.nyu.edu/jhaidt/disgust.scale.original.doc, accessed April 12, 2014.

18. Florian van Leeuwen et al., "Disgust Sensitivity Relates to Moral Foundations Independent of Political Ideology," *Evolutionary Behavioral Sciences* 11, no. 1 (June 2016): 92–98.

19. Julia Elad-Strenger, Jutta Proch, and Thomas Kessler, "Is Disgust a 'Conservative' Emotion?," *Personality and Social Psychology Bulletin* 46, no. 6 (October 2019): 896–912.

Chapter Fifteen: *Do Humans Dream of Electric Sheep?*

1. E. Tory Higgins et al., "Ideal Versus Ought Predilections for Approach and Avoidance: Distinct Self-Regulatory Systems," *Journal of Personality and Social Psychology* 66, no. 2 (February 1994): 276–86.

2. Lisa Feldman Barrett, *How Emotions Are Made: The Secret Life of the Brain* (New York: Houghton Mifflin Harcourt, 2017), 1.

3. James A. Coan Jr., "Lisa Feldman Barrett Bonus Material," in *Circle of Willis* podcast, MP3 audio, 24 minutes, http://circleofwillispodcast.com/episode/5a542 806ea4a43a8/lisa-feldman-barrett-bonus-material, accessed September 16, 2018.

4. James A. Russell, Jo-Anne Bachorowski, and José-Miguel Fernández-Dols, "Facial and Vocal Expressions of Emotion," *Annual Review of Psychology* 54, no. 1 (February 2003): 329–49.

5. *The Vault*, http://thevaultgame.com/, accessed September 20, 2018.

6. Thomas Dixon, email message to author, October 28, 2018.

7. Lisa Feldman Barrett, "Emotions Are Real," *Emotion* 12, no. 3 (June 2012): 413–29.

8. Check out Antonio Damasio, *The Feeling of What Happens: Body, Emotion and the Making of Consciousness* (London: Vintage, 2000); Antonio Damasio, *Looking for Spinoza: Joy, Sorrow and the Feeling Brain* (London: Vintage, 2004); Antonio Damasio, *Descartes' Error: Emotion, Reason and the Human Brain* (London: Vintage, 2006); Antonio Damasio, *The Strange Order of Things: Life, Feeling, and the Making of Cultures* (London: Random House, 2019).

9. A Bresó et al., "Usability and Acceptability Assessment of an Empathic Virtual Agent to Prevent Major Depression," *Expert Systems* 33, no. 4 (August 2016): 297–312.

10. See Affectiva website, http://go.affectiva.com/auto.

11. Lubomír Štěpánek, Jan Měšťák, and Pavel Kasal, "Machine-Learning at the Service of Plastic Surgery: A Case Study Evaluating Facial Attractiveness and Emotions Using R Language," *Proceedings of the Federated Conference on Computer Science and Information Systems* (2019): 107–12.

12. Bob Marcotte, "Using Data Science to Tell Which of These People Is Lying," University of Rochester Newscenter, May 22, 2018, https://www.rochester.edu /newscenter/data-science-facial-expressions-who-if-lying-321252/, accessed May 30, 2019.

13 Josh Chin, "Chinese Police Add Facial-Recognition Glasses to Surveillance Arsenal," *Wall Street Journal*, February 7, 2018, https://www.wsj.com/articles /chinese-police-go-robocop-with-facial-recognition-glasses-1518004353, accessed March 3, 2019.

14. The sequel, *Inside Out 2*, looks like it's adding eleven more emotions—boredom, tranquility, trust, embarrassment, crazy, jealousy, energetic, anxiety, mankini (?), genius, and kindness, but these bear no relation to any science I know of. Should be fun, though.

15. James A. Russell, "Core Affect and the Psychological Construction of Emotion," *Psychological Review* 110, no. 1 (January 2003): 145–72.

16. Gwyn Topham, "The End of Road Rage? A Car Which Detects Emotions," *The Guardian*, January 23, 2018, https://www.theguardian.com/business/2018/jan /23/a-car-which-detects-emotions-how-driving-one-made-us-feel, accessed September 20, 2018.

17. Lisa Feldman Barrett, "Can Machines Perceive Emotion?," Talks at Google, May 24, 2018, https://youtu.be/HlJQXfL_GeM.

18. E. F. Loftus and J. C. Palmer, "Eyewitness Testimony," in *Introducing Psychological Research: Sixty Studies That Shape Psychology,* ed. Philip Banyard and Andrew Grayson (London: Palgrave, 1996), 305–9.
19. Barrett, "Can Machines Perceive Emotion?"
20. Barrett, "Emotions Are Real," 418.

Epilogue: *The Last Feelings?*

1. It says, roughly: "One small step for man, one giant leap for the world." Close enough.
2. Kohske Takahashi, Takanori Oishi, and Masaki Shimada, "Is ☺ Smiling? Cross-Cultural Study on Recognition of Emoticon's Emotion," *Journal of Cross-Cultural Psychology* 48, no. 10 (November 2017): 1578–86.
3. Qiaozhu Mei, "Decoding the New World Language: Analyzing the Popularity, Roles, and Utility of Emojis," in *Companion Proceedings of the 2019 World Wide Web Conference,* ed. Ling Liu and Ryen White (New York: Association for Computing Machinery, 2019), 417–18; Hamza Alshenqeeti, "Are Emojis Creating a New or Old Visual Language for New Generations? A Socio-semiotic Study," *Advances in Language and Literary Studies* 7, no. 6 (December 2016): 56–69.
4. Ekman's original basic emotions: happiness, anger, sadness, disgust, surprise, and fear.
5. Don't get me started on the robots! Unless you want me to get started on the robots, that is.

Index

313

Index